START SMART
Back to School Refresher

Problem Solving

The Forests

Did You Know?
Many animals and plants live in the forests.

There are 10 🦋. 2 fly away. How many 🦋 are left?

Read

Look at the picture.
What do I already know? _____ 🦋 in all. _____ 🦋 fly away.

What do I need to find out? _____

Plan

I can write a number sentence.

Solve

The number sentence shows the difference.

_____ − _____ = _____ 🦋. _____ 🦋 are left.

Look Back

Does my answer make sense? _____   yes   no

Number Sense, Concepts, and Operations

Beautiful Birds

Did You Know?
You can see many beautiful birds around you.

Count on to add.

1.

   $4 + 3 =$ _____

2.

   $7 + 3 =$ _____

Use doubles plus 1 to find the sum.

3.

   $2 + 3 =$ _____

Quick Check

NA 3

Count back to subtract.

6 − 3 = _____

8 − 2 = _____

Use doubles to subtract.

12 − 6 = _____

Bird Problems

7. Use the picture. Write a problem. Solve your problem.

_____

_____

Name_____

Algebraic Thinking — Fall Leaves

Did You Know?
Some leaves change color in the fall.

Leaf Patterns

1.

How many kinds of leaves are in the pattern? _____

2. Circle the leaf that comes next in the pattern above.

A Different Pattern

③ Circle the pattern unit.

④ Circle the leaf that comes next in the pattern.

Leaf Patterns

⑤ ✏️ Write **About It!** Use two different color crayons to make a color pattern. Have others describe your pattern. Use letters to show your pattern another way.

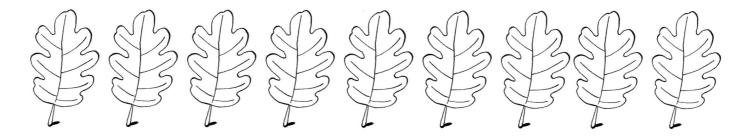

___ ___ ___ ___ ___ ___ ___ ___ ___

Data Analysis and Probability

At School

Did You Know? School children in the United States share different cultures.

Boys and Girls • Picture Graphs

Look at the picture graph.

It shows how many boys and girls are in Mrs. Bright's class.

Children in Mrs. Bright's Class	
Boys	☺ ☺ ☺ ☺ ☺ ☺ ☺
Girls	☺ ☺ ☺ ☺ ☺ ☺ ☺ ☺ ☺

Key: Each ☺ stands for 1 child.

1. How many children does each ☺ stand for? _____

2. How many girls are in Mrs. Bright's class? _____

3. Are there more boys or girls in Mrs. Bright's class?

_____

Family Pets • Tally Charts

Look at the tally chart.
Use facts from the tally chart for problems 4–6.

Have Pets	Have No Pets
卌 卌	卌 l

④ How many families have pets? _____

⑤ How many families do not have pets? _____

⑥ How many more families have pets than families who do not? _____

Tell how you know. _____

_____

Make a Tally Chart

⑦ How many children in your class have pets? How many children in your class do not have pets? Collect the facts. Show what you find out. Make a tally chart.

Have Pets	Have No Pets

⑧ **Write About It!** Write a sentence that tells about your tally chart.

_____

_____

Name_____

Measurement

State Flag of Alaska

Did You Know?
Every state has its own flag.

Flag Measures • Length

1. The picture of the flag is about 2 📎 high.

   About how many 📎 wide is it?

   about _____ 📎 wide

2. About how many 📎 high is this picture

   of the flag pole? about _____ 📎 high

3. Would you use inches or feet to measure this

   picture of the flag? _____

4. Would you use inches or feet to measure the real flag?

   _____

Quick Check

NA 9

Animal Measures • Inches or Feet?

5 Cougars are big and fast.

Would you use inches or feet to measure the length of the real cougar? _____

6 Mockingbirds are small but they can sing sweetly.

Would you use inches or feet to measure the length of the real bird? _____

Classroom Measure

7 Write **About It!** Find something in your classroom you would measure in feet. Draw a picture.

Name_____

Geometry and Spatial Sense

At the Beach

Did You Know?
The United States has many beautiful beaches. A day at the beach can be fun.

What Object? • 3-Dimensional Figures

1. Circle the object that has the same figure as the .

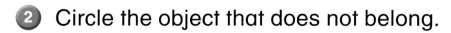

2. Circle the object that does not belong.

Quick Check

NA 11

From 3 Dimensions to 2 Dimensions

3 Circle the shape you would make if you traced the CRACKERS box.

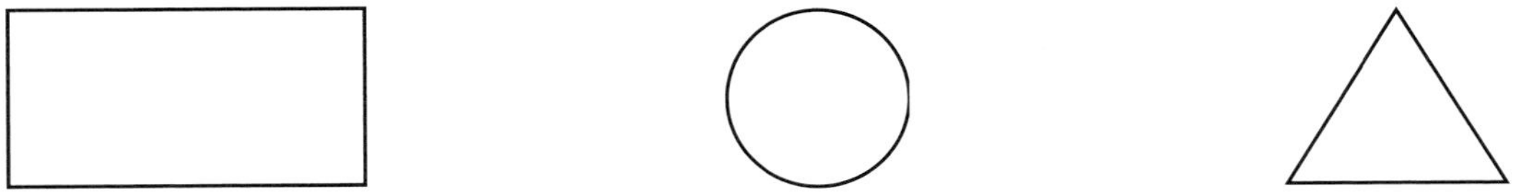

What Shape? • 2-Dimensional Shapes
How many sides and vertices does each shape have?

4

Shape	How many sides?	How many vertices?
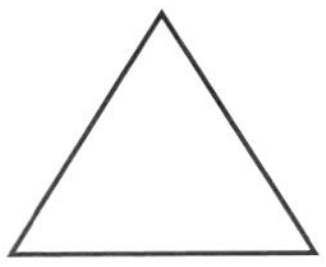 △		
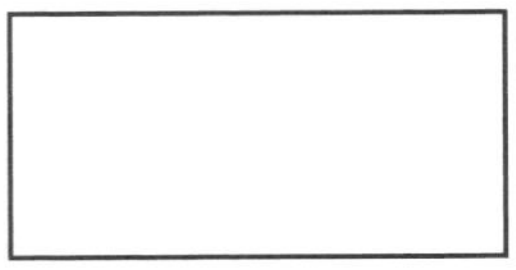 ▭		
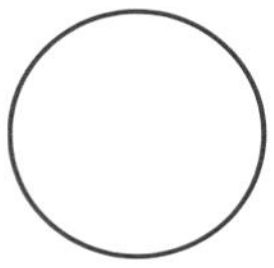 ○		

5 Circle the shape that does not belong. Tell how you know.

_____

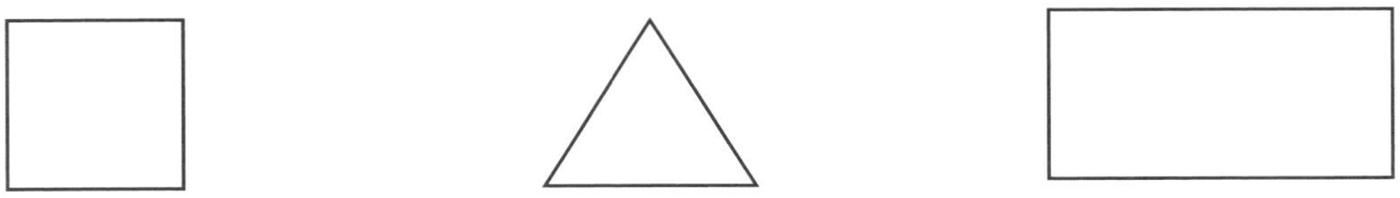

6 ✏️ **Write About It!** Circle the shape that does not belong. Tell how you know.

_____

NA 12

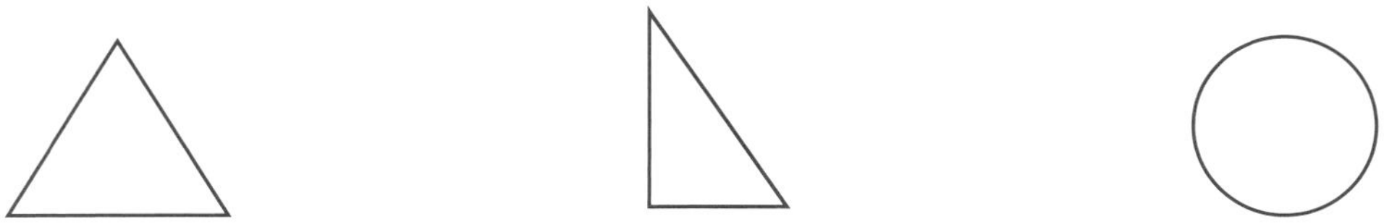

MACMILLAN/McGRAW-HILL
Math

Macmillan
McGraw-Hill

PROGRAM AUTHORS

Douglas H. Clements, Ph.D.
Professor of Mathematics Education
State University of
  New York at Buffalo
Buffalo, New York

Carol E. Malloy, Ph.D.
Assistant Professor of
  Mathematics Education
University of North Carolina at
  Chapel Hill
Chapel Hill, North Carolina

Lois Gordon Moseley
Mathematics Consultant
Houston, Texas

Yuria Orihuela
District Math Supervisor
Miami-Dade County Public Schools
Miami, Florida

Robyn R. Silbey
Montgomery County Public Schools
Rockville, Maryland

SENIOR CONTENT REVIEWERS

Gunnar Carlsson, Ph.D.
Professor of Mathematics
Stanford University
Stanford, California

Ralph L. Cohen, Ph.D.
Professor of Mathematics
Stanford University
Stanford, California

Macmillan
McGraw-Hill

Published by Macmillan/McGraw-Hill, of McGraw-Hill Education, a division of The McGraw-Hill Companies, Inc., Two Penn Plaza, New York, New York 10121.

Copyright © 2005 by Macmillan/McGraw-Hill. All rights reserved. No part of this publication may be reproduced or distributed in any form or by any means, or stored in a database or retrieval system, without the prior written consent of The McGraw-Hill Companies, Inc., including, but not limited to, network storage or transmission, or broadcast for distance learning.

Foldables™, Math Tool Chest™, Math Traveler™, Mathematics Yes!™, Yearly Progress Pro™, and Math Background for Professional Development™ are trademarks of The McGraw-Hill Companies, Inc.

Printed in the United States of America

ISBN 0-02-104003-6/2
3 4 5 6 7 8 9 073 08 07 06 05 04

RFB&D
learning through listening

Students with print disabilities may be eligible to obtain an accessible, audio version of the pupil edition of this textbook. Please call Recording for the Blind & Dyslexic at 1-800-221-4792 for complete information.

CONTRIBUTING AUTHORS

Mary Behr Altieri
1993 Presidential Awardee
Putnam/Northern Westchester BOCES
Yorktown Heights, New York

Ellen C. Grace
Educational Consultant
Albuquerque, New Mexico

Dinah Zike
Dinah Might Adventures
Comfort, Texas

CONSULTANTS

ASSESSMENT

**Lynn Fuchs, Ph.D.,
and Douglas Fuchs, Ph.D.**
Department of Special Education
Vanderbilt University
Nashville, Tennessee

PROFESSIONAL DEVELOPMENT

Nadine Bezuk, Ph.D.
Director, School of Teacher Education
San Diego State University
San Diego, California

READING AND MATH

Karen D. Wood, Ph.D.
Professor, Dept. of Reading and
  Elementary Education
University of North Carolina
  at Charlotte
Charlotte, North Carolina

ESL

Sally S. Blake, Ph.D.
Associate Professor,
  Teacher Education
The University of Texas at El Paso
El Paso, Texas

Josefina Villamil Tinajero, Ed.D.
Professor of Bilingual Education
Interim Dean, College of Education
The University of Texas at El Paso
El Paso, Texas

Contents

UNIT 1
Chapters 1-2

Numbers and Addition and Subtraction Facts

CHAPTER 1 THEME: Around the Pond

MATH STORY: By the Pond

1 EXPLORING NUMBER RELATIONSHIPS 1
Math at Home 2
1 HANDS ON Numbers 3
2 ALGEBRA: Number Patterns x 5
3 HANDS ON Understand Addition and Subtraction 7
4 PROBLEM SOLVING SKILL: READING FOR MATH 9
PROBLEM SOLVING: PRACTICE 11
✏️ Writing for Math 12
Chapter Review/Test 13
Spiral Review and Test Prep 14

CHAPTER 2 THEME: Class Trip

MATH POEM: Class Trip

2 ADDITION STRATEGIES AND FACTS TO 20 15
Math at Home 16
1 ALGEBRA: Order Property and Zero Property x 17
2 HANDS ON Count On to Add 19
3 ALGEBRA: Addition Patterns x 21
4 ALGEBRA: Add Three Numbers x 23
5 PROBLEM SOLVING: STRATEGY Draw a Picture 25
Game Zone 27
Technology Link: Calculator 28
Chapter Review/Test 29
Spiral Review and Test Prep 30

LOG ON
Activities referenced on pp. 2, 14, 16, 30, 32, 40, 50, 52, 62, 68

e-Journal pp. 12, 48, 73

www.mmhmath.com

Technology
Math Traveler, p. 66
Math Tool Chest, p. 66
Multimedia Glossary, pp. G1–G15

iv

UNIT 1
Chapters 3-4

Numbers and Addition and Subtraction Facts

CHAPTER 3 THEME: On the Go

FINGER PLAY: Twelve Little Rabbits

3 SUBTRACTION STRATEGIES AND FACTS TO 20 31
   Math at Home 32
1 Count Back to Subtract 33
2 HANDS ON Subtract All and Subtract Zero 35
3 HANDS ON/ALGEBRA: Relate Addition to Subtraction x 37
   Extra Practice 39
4 ALGEBRA: Missing Number x 41
5 ALGEBRA: Names for Numbers x 43
6 PROBLEM SOLVING SKILL: READING FOR MATH 45
   PROBLEM SOLVING: PRACTICE 47
      Writing for Math 48
   Chapter Review/Test 49
   Spiral Review and Test Prep 50

CHAPTER 4 THEME: Nature Walk

MATH SONG: Bubbles

4 EXPLORING ADDITION AND SUBTRACTION 51
   Math at Home 52
1 Use Doubles to Add and Subtract 53
2 HANDS ON Use 10 to Add and Subtract 9 55
3 HANDS ON Use 10 to Add and Subtract 7 and 8 57
4 Fact Families 59
   Extra Practice 61
5 ALGEBRA: PROBLEM SOLVING: STRATEGY Write a Number Sentence x 63
   Game Zone 65
   Technology Link: Computer 66
   Chapter Review/Test 67
   Spiral Review and Test Prep 68

UNIT REVIEW
   TIME Time for Kids 68A
   PROBLEM SOLVING: DECISION MAKING 69
   Study Guide and Review 71
   Performance Assessment 73
   Enrichment: Missing Numbers 74

HANDS ON Lessons and Activities in Every Unit

x ALGEBRA

v

UNIT 2
Chapters 5-6

Place Value and Money

CHAPTER 5 THEME: Sing and Dance
MATH STORY: Snap and Tap!

5 UNDERSTANDING PLACE VALUE 75
Math at Home 76
1 HANDS ON Tens 77
2 HANDS ON Tens and Ones 79
3 Place Value to 100 81
4 Read and Write Numbers 83
5 Estimate Numbers 85
6 PROBLEM SOLVING SKILL: READING FOR MATH 87
PROBLEM SOLVING: PRACTICE 89
Writing for Math 90
Chapter Review/Test 91
Spiral Review and Test Prep 92

CHAPTER 6 THEME: At the Park
MATH SONG: Skip-Count Song

6 NUMBERS AND PATTERNS 93
Math at Home 94
1 HANDS ON/ALGEBRA: Compare Numbers x 95
2 Order Numbers 97
3 ALGEBRA: Skip-Counting Patterns x 99
Extra Practice 101
4 Ordinal Numbers 103
5 HANDS ON Even and Odd Numbers 105
6 PROBLEM SOLVING: STRATEGY Use Logical Reasoning 107
Game Zone 109
Technology Link: Calculator 110
Chapter Review/Test 111
Spiral Review and Test Prep 112

LOG ON
Activities referenced on pp. 76, 92, 94, 102, 112, 114, 130, 132, 146

e-Journal pp. 90, 128, 151

www.mmhmath.com

Technology
Math Traveler, p. 144
Math Tool Chest, p. 144
Multimedia Glossary, pp. G1–G15

UNIT 2
Chapters 7-8

Place Value and Money

CHAPTER 7 THEME: At the Fair

MATH POEM: A Visit to the Fair

7 MONEY 113
Math at Home 114
1 HANDS ON Pennies, Nickels, and Dimes 115
2 HANDS ON Count Coin Collections 117
3 HANDS ON Money and Place Value 119
4 Quarters and Half Dollars 121
5 HANDS ON Make Equal Amounts 123
6 PROBLEM SOLVING SKILL: READING FOR MATH 125
PROBLEM SOLVING: PRACTICE 127
Writing for Math 128
Chapter Review/Test 129
Spiral Review and Test Prep 130

CHAPTER 8 THEME: Save and Spend

MATH POEM: Money Rhymes

8 USING MONEY 131
Math at Home 132
1 HANDS ON Dollar 133
2 Dollars and Cents 135
3 Compare Money Amounts 137
4 HANDS ON Make Change 139
5 PROBLEM SOLVING: STRATEGY Act It Out 141
Game Zone 143
Technology Link: Computer 144
Chapter Review/Test 145
Spiral Review and Test Prep 146

UNIT REVIEW
Time for Kids 146A
PROBLEM SOLVING: LINKING MATH AND SCIENCE 147
Study Guide and Review 149
Performance Assessment 151
Enrichment: Different Ways to Show Numbers 152

Lessons and Activities in Every Unit

ALGEBRA

vii

UNIT 3
Chapters 9-10

Time, Graphs, and Regrouping

CHAPTER 9 THEME: On Time

MATH STORY: Let the Show Begin

9 TELLING TIME 153
Math at Home 154
1 **HANDS ON** Time to the Hour and Half Hour 155
2 Time to Five Minutes 157
3 Time to the Quarter Hour 159
4 Time Before and After the Hour 161
5 **PROBLEM SOLVING SKILL: READING FOR MATH** 163
   PROBLEM SOLVING: PRACTICE 165
   Writing for Math 166
   Chapter Review/Test 167
   Spiral Review and Test Prep 168

CHAPTER 10 THEME: Around the Clock

MATH FINGERPLAY: Going to Bed

10 TIME AND CALENDAR 169
Math at Home 170
1 A.M. and P.M. 171
2 **HANDS ON** Elapsed Time 173
   Extra Practice 175
3 Calendar 177
4 **ALGEBRA:** Time Relationships 179
5 **PROBLEM SOLVING: STRATEGY** Use a Model 181
   Game Zone 183
   Technology Link: Calculator 184
   Chapter Review/Test 185
   Spiral Review and Test Prep 186

LOG ON
Activities referenced on pp. 154, 168, 170, 176, 186, 188, 206, 208, 222

e-Journal pp. 166, 204, 227

www.mmhmath.com

Technology
Math Traveler, p. 220
Math Tool Chest, p. 220
Multimedia Glossary, pp. G1–G15

viii

UNIT 3
Chapters 11-12

Time, Graphs, and Regrouping

CHAPTER 11 THEME: Rain or Shine

MATH POEM: Under the Weather

11 DATA AND GRAPHS 187
Math at Home 188
1 Picture and Bar Graphs 189
2 HANDS ON Surveys 191
3 Make a Bar Graph 193
4 Pictographs 195
5 Line Plots 197
6 Different Ways to Show Data 199
7 PROBLEM SOLVING SKILL: READING FOR MATH 201
   PROBLEM SOLVING: PRACTICE 203
   Writing for Math 204
   Chapter Review/Test 205
   Spiral Review and Test Prep 206

CHAPTER 12 THEME: Bake Sale

MATH POEM: Bushy-Tailed Mathematicians

12 NUMBER RELATIONSHIPS AND REGROUPING 207
Math at Home 208
1 HANDS ON Explore Regrouping 209
2 HANDS ON Addition with Sums to 20 211
3 HANDS ON Addition with Greater Numbers 213
4 ALGEBRA: Renaming Numbers x 215
5 PROBLEM SOLVING: STRATEGY Use a Pattern 217
   Game Zone 219
   Technology Link: Computer 220
   Chapter Review/Test 221
   Spiral Review and Test Prep 222

UNIT REVIEW
Time for Kids 222A
PROBLEM SOLVING: LINKING MATH AND SCIENCE 223
Study Guide and Review 225
Performance Assessment 227
Enrichment: Venn Diagrams 228

Lessons and Activities in Every Unit

x ALGEBRA

ix

UNIT 4
Chapters 13–14

Add and Subtract 2-Digit Numbers

CHAPTER 13 THEME: On the Job
MATH STORY: Side by Side

13 2-DIGIT ADDITION 229
  Math at Home 230
  1 HANDS ON Mental Math: Add Tens 231
     Extra Practice 233
  2 Count on Tens and Ones to Add 235
  3 HANDS ON Decide When to Regroup 237
  4 Add a 1-Digit and a 2-Digit Number 239
  5 Add 2-Digit Numbers 241
  6 HANDS ON Practice Addition 243
  7 PROBLEM SOLVING SKILL: READING FOR MATH 245
     PROBLEM SOLVING: PRACTICE 247
     Writing for Math 248
     Chapter Review/Test 249
     Spiral Review and Test Prep 250

CHAPTER 14 THEME: Pet Pals
MATH POEM: No Room in the Tub

14 PRACTICE AND APPLY 2-DIGIT ADDITION 251
  Math at Home 252
  1 Rewrite Addition 253
  2 Practice 2-Digit Addition 255
  3 ALGEBRA: Check Addition x 257
  4 HANDS ON Estimate Sums 259
  5 ALGEBRA: Add Three Numbers x 261
  6 PROBLEM SOLVING: STRATEGY Choose a Method ... 263
     Game Zone 265
     Technology Link: Calculator 266
     Chapter Review/Test 267
     Spiral Review and Test Prep 268

LOG ON
Activities referenced on pp. 230, 234, 250, 252, 268, 270, 274, 290, 292, 308

e-Journal pp. 248, 288, 313

www.mmhmath.com

Technology
Math Traveler, p. 306
Math Tool Chest, p. 306
Multimedia Glossary, pp. G1–G15

UNIT 4
Chapters 15-16

Add and Subtract 2-Digit Numbers

CHAPTER 15 THEME: Party Time

MATH POEM: Tea Party

15 2-DIGIT SUBTRACTION 269
    Math at Home 270
1 Mental Math: Subtract Tens 271
    Extra Practice 273
2 Mental Math: Count Back Tens and Ones to Subtract 275
3 **HANDS ON** Decide When to Regroup 277
4 **HANDS ON** Subtract a 1-Digit Number from a 2-Digit Number 279
5 **HANDS ON** Subtract 2-Digit Numbers 281
6 **HANDS ON** Practice Subtraction 283
7 **PROBLEM SOLVING SKILL: READING FOR MATH** 285
    PROBLEM SOLVING: PRACTICE 287
    Writing for Math 288
    Chapter Review/Test 289
    Spiral Review and Test Prep 290

CHAPTER 16 THEME: Family Fun

MATH SONG: Pancake Song

16 PRACTICE AND APPLY 2-DIGIT SUBTRACTION 291
    Math at Home 292
1 Rewrite 2-Digit Subtraction 293
2 Practice 2-Digit Subtraction 295
3 **ALGEBRA:** Check Subtraction x 297
4 **HANDS ON** Estimate Differences 299
5 **ALGEBRA:** Mental Math: Strategies x 301
6 **ALGEBRA: PROBLEM SOLVING: STRATEGY** Choose the Operation x ... 303
    Game Zone 305
    Technology Link: Computer 306
    Chapter Review/Test 307
    Spiral Review and Test Prep 308

UNIT REVIEW
    TIME Time for Kids 308A
PROBLEM SOLVING: DECISION MAKING 309
Study Guide and Review 311
Performance Assessment 313
Enrichment: Solve a Simple Equation 314

Lessons and Activities in Every Unit

UNIT 5
Chapters 17-18

Measurement and Geometry

CHAPTER 17 THEME: Long Ago

MATH STORY: Animal Tracks and Footprints

17 ESTIMATE AND MEASURE LENGTH 315
- Math at Home 316
- 1 HANDS ON Nonstandard Units of Length 317
- 2 HANDS ON Measure to the Nearest Inch 319
- 3 HANDS ON Inch, Foot, and Yard 321
- 4 HANDS ON Centimeter and Meter 323
- 5 PROBLEM SOLVING SKILL: READING FOR MATH 325
  - PROBLEM SOLVING: PRACTICE 327
  - Writing for Math 328
- Chapter Review/Test 329
- Spiral Review and Test Prep 330

CHAPTER 18 THEME: Fill It Up

MATH SONG: Measuring Song

18 ESTIMATE AND MEASURE CAPACITY AND WEIGHT ... 331
- Math at Home 332
- 1 HANDS ON Explore Capacity 333
- 2 HANDS ON/ALGEBRA: Fluid Ounce, Cup, Pint, Quart, and Gallon 𝒳 335
- 3 HANDS ON Ounce and Pound 337
- 4 HANDS ON Milliliter and Liter 339
- 5 HANDS ON Gram and Kilogram 341
- 6 Temperature 343
- 7 PROBLEM SOLVING: STRATEGY Use Logical Reasoning 345
- Game Zone 347
- Technology Link: Calculator 348
- Chapter Review/Test 349
- Spiral Review and Test Prep 350

LOG ON
Activities referenced on pp. 316, 330, 332, 350, 352, 362, 370, 372, 388

e-Journal pp. 328, 368, 393

www.mmhmath.com

Technology
Math Traveler, p. 386
Math Tool Chest, p. 386
Multimedia Glossary, pp. G1–G15

UNIT 5
Chapters 19-20

Measurement and Geometry

CHAPTER 19 THEME: Welcome Home
MATH POEM: A Riddle

19 2-DIMENSIONAL AND 3-DIMENSIONAL SHAPES 351
    Math at Home 352
1. 3-Dimensional Figures 353
2. 2-Dimensional Shapes 355
3. HANDS ON 2-Dimensional and 3-Dimensional Relationships 357
4. HANDS ON Combine Shapes 359
    Extra Practice 361
5. HANDS ON/ALGEBRA: Shape Patterns x 363
6. PROBLEM SOLVING SKILL: READING FOR MATH 365
    PROBLEM SOLVING: PRACTICE 367
    Writing for Math 368
    Chapter Review/Test 369
    Spiral Review and Test Prep 370

CHAPTER 20 THEME: Fun City
MATH POEM: Paper Dolls

20 SPATIAL SENSE 371
    Math at Home 372
1. HANDS ON Congruence 373
2. HANDS ON Symmetry 375
3. HANDS ON Slides, Flips, and Turns 377
4. HANDS ON Perimeter 379
5. HANDS ON Area 381
6. ALGEBRA: PROBLEM SOLVING: STRATEGY Guess and Check x 383
    Game Zone 385
    Technology Link: Computer 386
    Chapter Review/Test 387
    Spiral Review and Test Prep 388

UNIT REVIEW
    TIME Time for Kids 388A
    PROBLEM SOLVING: LINKING MATH AND SCIENCE 389
    Study Guide and Review 391
    Performance Assessment 393
    Enrichment: Volume 394

HANDS ON Lessons and Activities in Every Unit

x ALGEBRA

UNIT 6
Chapters 21-22

Understanding Greater Numbers

CHAPTER 21 THEME: Collections

MATH STORY: We Can Make Almost Anything

21 PLACE VALUE TO THOUSANDS 395
- Math at Home 396
- 1 HANDS ON Hundreds 397
- 2 Hundreds, Tens, and Ones 399
- 3 Place Value Through Hundreds 401
- Extra Practice 403
- 4 HANDS ON Explore Place Value To Thousands 405
- 5 PROBLEM SOLVING SKILL: READING FOR MATH 407
- PROBLEM SOLVING: PRACTICE 409
- Writing for Math 410
- Chapter Review/Test 411
- Spiral Review and Test Prep 412

CHAPTER 22 THEME: Craft Show

MATH SONG: Pattern Song

22 NUMBER RELATIONSHIPS AND PATTERNS 413
- Math at Home 414
- 1 ALGEBRA: Compare Numbers 415
- 2 Order Numbers on a Number Line 417
- 3 Order Numbers 419
- 4 ALGEBRA: Number Patterns 421
- 5 Count Forward, Count Backward 423
- 6 ALGEBRA: PROBLEM SOLVING: STRATEGY Make a Table 425
- Game Zone 427
- Technology Link: Computer 428
- Chapter Review/Test 429
- Spiral Review and Test Prep 430

LOG ON
Activities referenced on pp. 396, 404, 412, 414, 430, 432, 436, 446, 448, 452, 464

e-Journal pp. 410, 444, 469

www.mmhmath.com

Technology
Math Traveler, p. 428
Math Tool Chest, p. 428
Multimedia Glossary, pp. G1–G15

UNIT 6
Chapters 23-24

Understanding Greater Numbers

CHAPTER 23 THEME: At the Circus

MATH POEM: Main Attraction

23 3-DIGIT ADDITION . . . 431
    Math at Home . . . 432
1. Add Hundreds . . . 433
    Extra Practice . . . 435
2. **HANDS ON** Regroup Ones . . . 437
3. **HANDS ON** Regroup Tens . . . 439
4. PROBLEM SOLVING SKILL: READING FOR MATH . . . 441
    PROBLEM SOLVING: PRACTICE . . . 443
    Writing for Math . . . 444
    Chapter Review/Test . . . 445
    Spiral Review and Test Prep . . . 446

CHAPTER 24 THEME: On Vacation

MATH SONG: Sea Song

24 3-DIGIT SUBTRACTION . . . 447
    Math at Home . . . 448
1. Subtract Hundreds . . . 449
    Extra Practice . . . 451
2. **HANDS ON** Regroup Tens as Ones . . . 453
3. **HANDS ON** Regroup Hundreds as Tens . . . 455
4. Estimate, Add, and Subtract Money Amounts . . . 457
5. PROBLEM SOLVING: STRATEGY Work Backward . . . 459
    Game Zone . . . 461
    Technology Link: Calculator . . . 462
    Chapter Review/Test . . . 463
    Spiral Review and Test Prep . . . 464

UNIT REVIEW
Time for Kids . . . 464A
PROBLEM SOLVING: LINKING MATH AND SCIENCE . . . 465
Study Guide and Review . . . 467
Performance Assessment . . . 469
Enrichment: Adding and Subtracting Money . . . 470

Lessons and Activities in Every Unit

UNIT 7
Chapters 25-26

Fractions, Probability, Data, and Operations

CHAPTER 25 THEME:
Deep Blue Sea

MATH STORY:
Under the Sea

25 FRACTIONS 471
Math at Home 472
1 Unit Fractions 473
2 Fractions Equal to 1 475
3 Other Fractions 477
4 HANDS ON Unit Fractions of a Group 479
5 Other Fractions of a Group 481
6 ALGEBRA: Compare Fractions x 483
7 PROBLEM SOLVING SKILL: READING FOR MATH 485
  PROBLEM SOLVING: PRACTICE 487
  Writing for Math 488
  Chapter Review/Test 489
  Spiral Review and Test Prep 490

CHAPTER 26 THEME:
What Are the Chances?

MATH POEM:
Heads or Tails?

26 PROBABILITY 491
Math at Home 492
1 Explore Probability 493
2 HANDS ON More Likely, Equally Likely, or Less Likely 495
3 HANDS ON Make Predictions 497
  Extra Practice 499
4 PROBLEM SOLVING: STRATEGY Make a List 501
  Game Zone 503
  Technology Link: Calculator 504
  Chapter Review/Test 505
  Spiral Review and Test Prep 506

LOG ON
Activities referenced on pp. 472, 490, 492, 500, 506, 508, 522, 524, 532, 542

e-Journal   pp. 488, 520, 547

www.mmhmath.com

Technology
Math Traveler, p. 540
Math Tool Chest, p. 540
Multimedia Glossary, pp. G1–G15

UNIT 7
Chapters 27-28

Fractions, Probability, Data, and Operations

CHAPTER 27 THEME: All About Us

MATH POEM: The Giraffe Graph

27 INTERPRETING DATA 507
Math at Home 508
1 Range and Mode 509
2 HANDS ON Median 511
3 ALGEBRA: Coordinate Graphs x 513
4 HANDS ON Line Graphs 515
5 PROBLEM SOLVING SKILL: READING FOR MATH 517
PROBLEM SOLVING: PRACTICE 519
✏️ Writing for Math 520
Chapter Review/Test 521
Spiral Review and Test Prep 522

CHAPTER 28 THEME: Bugs, Bugs, Bugs

MATH POEM: Caterpillar Pete

28 EXPLORING MULTIPLICATION AND DIVISION 523
Math at Home 524
1 HANDS ON Explore Equal Groups 525
2 HANDS ON Repeated Addition and Multiplication 527
3 HANDS ON/ALGEBRA: Use Arrays to Multiply x 529
Extra Practice 531
4 HANDS ON Repeated Subtraction and Division 533
5 HANDS ON Divide to Find Equal Shares 535
6 ALGEBRA: PROBLEM SOLVING: STRATEGY Use a Pattern x 537
Game Zone 539
Technology Link: Computer 540
Chapter Review/Test 541
Spiral Review and Test Prep 542

UNIT REVIEW
TIME Time for Kids 542A
PROBLEM SOLVING: DECISION MAKING 543
Study Guide and Review 545
Performance Assessment 547
Enrichment: Remainder 548

Picture Glossary G1

HANDS ON
Lessons and Activities in Every Unit

x ALGEBRA

xvii

Amazing Math!

The Foot Fits

Measure the length of your foot.
Then measure from your wrist to your elbow.
What do you notice?

_____

_____

Learn more about measurement in Chapter 17.

Lullaby Lion

A lion sleeps about 20 hours a day! How many hours do you sleep each day?

_____ hours

How many more hours of sleep does a lion get than you?

_____ hours

Learn more about subtraction in Chapters 3 and 15.

UNIT 1
CHAPTER 1

Exploring Number Relationships

By the Pond

Story by Becky Manfredini
Illustrated by Bernard Adnet

4 snails sit on a log by a gate.
4 more join them.

Now there are _____.

10 sunfish swim and do some tricks.
4 hide.

Now there are _____.

8 frogs leap and leap again.
2 more join them.

Now there are _____.

12 crayfish creep on the blue pond's floor.
8 go and hide.

Now there are _____.

Math at Home

Dear Family,

I will review numbers and facts to 12 in Chapter 1. Here are my math words and an activity that we can do together.

Love, _____

My Math Words

skip-count :

2, 4, 6, 8, 10, 12

sum :

$$\begin{array}{r} 4 \\ +3 \\ \hline 7 \end{array}$$ ← sum

difference :

$$\begin{array}{r} 8 \\ -2 \\ \hline 6 \end{array}$$ ← difference

Home Activity

Ask your child questions that require a number in the answer. Example questions might be, "How many windows are in this room?" or "How many crackers did you eat today?" Have your child respond to these questions. Switch roles and repeat.

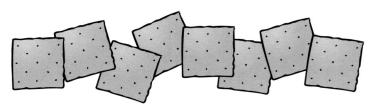

Books to Read

Look for these books at your local library and use them to help your child explore number relationships.

- **Monster Math** by Grace Maccarone, Scholastic, 1995.
- **12 Ways To Get to 11** by Eve Merriam, Simon & Schuster, 1993.
- **So Many Cats!** By Beatrice Schenk de Regniers, Clarion Books, 1985.

www.mmhmath.com
For Real World Math Activities

Name _____ **Numbers**

Learn You can use numbers in different ways.
You can use numbers to count.
How many  are there?

____4____ 🐟

Your Turn Count these things in your classroom.
Write the number.

① How many are there? ② How many are there?

_____   _____

③ How many are there? ④ How many are there?

_____   _____

⑤ **Write About It!** Tell one way you used a number today.

Chapter 1 Lesson 1   three **3**

Practice Tell about yourself. Write the number.

Numbers are all around you!

6. Your age _____
7. Your address _____
8. Number of girls in your class _____
9. Number of boys in your class _____
10. Time school starts _____
11. Time school ends _____
12. Number of books in your desk _____
13. Number of pencils in your desk _____

Math at Home: Your child practiced counting objects.
Activity: Have your child tell how many objects, such as shoes or doors, are in a room.

Name_____

Understand Addition and Subtraction

Learn Add to find the sum.

6 turtles are on the rock.
4 more turtles get on the rock.
How many turtles are on the rock now?

6 + 4 = 10

This is an addition sentence.

10 turtles on the rock now

Math Words
add
sum
difference

Subtract to find the difference.

3 of the turtles slide off the rock.
How many turtles are left on the rock?

10 − 3 = 7

This is a subtraction sentence.

7 turtles left on the rock

Try It Add or subtract. You can draw a picture to help.

① 4 + 1 = 5

② 3 + 3 = ___

③ 7 − 2 = ___

④ 6 − 5 = ___

⑤ **Write About It!** How are addition and subtraction different?

Chapter 1   Lesson 3

Practice You can show addition and subtraction two ways.

⑥ $5 + 6 = \underline{11}$ $\begin{array}{r} 5 \\ +6 \\ \hline 11 \end{array}$ ⑦ $11 - 5 = \underline{6}$ $\begin{array}{r} 11 \\ -5 \\ \hline 6 \end{array}$

Add or subtract.

⑧ $\begin{array}{r}6\\+2\\\hline\end{array}$	⑨ $\begin{array}{r}5\\+3\\\hline\end{array}$	⑩ $\begin{array}{r}8\\+2\\\hline\end{array}$	⑪ $\begin{array}{r}7\\+2\\\hline\end{array}$	⑫ $\begin{array}{r}4\\+3\\\hline\end{array}$
⑬ $\begin{array}{r}5\\+2\\\hline\end{array}$	⑭ $\begin{array}{r}1\\+8\\\hline\end{array}$	⑮ $\begin{array}{r}4\\+6\\\hline\end{array}$	⑯ $\begin{array}{r}2\\+4\\\hline\end{array}$	⑰ $\begin{array}{r}2\\+9\\\hline\end{array}$
⑱ $\begin{array}{r}8\\-2\\\hline\end{array}$	⑲ $\begin{array}{r}9\\-1\\\hline\end{array}$	⑳ $\begin{array}{r}10\\-0\\\hline\end{array}$	㉑ $\begin{array}{r}12\\-7\\\hline\end{array}$	㉒ $\begin{array}{r}5\\-2\\\hline\end{array}$
㉓ $\begin{array}{r}6\\-1\\\hline\end{array}$	㉔ $\begin{array}{r}10\\-8\\\hline\end{array}$	㉕ $\begin{array}{r}11\\-6\\\hline\end{array}$	㉖ $\begin{array}{r}10\\-3\\\hline\end{array}$	㉗ $\begin{array}{r}4\\-0\\\hline\end{array}$

Problem Solving **Number Sense**

㉘ 5 birds sit in a tree.
3 birds fly away.
How many birds are left? _____ birds

Math at Home: Your child reviewed addition and subtraction facts to 12.
Activity: Have your child add 6 + 3, then subtract 10 − 5. Encourage him or her to draw a picture to represent each problem.

Name_____

Problem Solving Skill
Reading for Math

At the Pond

These animals share a pond.
4 ducks swim on the water.
6 turtles sit on logs.
Frogs rest on lily pads.

Reading Skill: Use Illustrations

1. How many ducks are on the pond? _____ ducks

2. There are 2 frogs on each lily pad.
   How many frogs are there?
   Skip-count by twos.
   Write the numbers. _____ _____ frogs

3. 2 of the turtles jump off the log.
   How many are left on the log? _____ turtles

4. 3 more ducks come to the pond.
   How many ducks are there in all? _____ ducks

Chapter 1   Lesson 4

Pond Life

A pond is full of water life.
4 beavers rest on the land.
7 fish swim under the water.
Cattails grow around the pond.

Use Illustrations

5. How many fish are in the pond? _____ fish

6. There are 5 cattails in each group.
   How many cattails are there?
   Skip-count by fives.
   Write the numbers. _____ _____ cattails

7. 2 more fish join the group of fish.
   How many fish are there? _____ fish

8. 2 beavers swim away.
   How many are left? _____ beavers

Math at Home: Your child used an illustration to solve problems.
Activity: Use a picture from a book or magazine. Ask questions about the picture, such as, "How many cats are there?" or "If two dogs walk away, how many are left?"

Name _____

Problem Solving Practice

Solve.

1) There are 5 on a lily pad. 3 more jump on.

How many are there in all? _____

How many frogs do you see on the lily pad? _____

Write the number of frogs that jump on. _____

Write a number sentence. _____ + _____ = _____

2) 9 turtles sit on a log. 4 turtles get hot and jump in the water. How many turtles are left on the log?

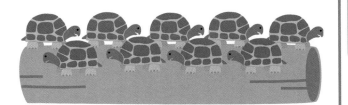

How many turtles do you see on the log? _____ turtles

Write the number of turtles that jump off. _____ turtles

Write a number sentence. _____ − _____ = _____ turtles

Write a Story!

3) Write a subtraction problem about this number sentence. Find the difference. $12 - 8 =$ _____

_____

Chapter 1 Problem Solving Practice

eleven **11**

Writing for Math

Write an addition problem about this picture. Use the words on the cards to help.

| duck | log |

Think

How many ducks are on the log? _____

How many ducks are in the water? _____

What addition fact do I see in the picture?

_____ + _____ = _____

Solve

I can write my problem now.

_____

_____

_____

Explain

I can tell you how my addition problem matches the addition fact.

_____

_____

Chapter 1
Review/Test

Name _____

1. Skip-count the bugs by twos. Write the number.

____ ____ ____ ____ ____ ____

2. Skip-count the bugs by fives. Write the number.

____ ____ ____ ____ ____ ____

Add or subtract.

3. 4 + 3 = ___	**4.** 9 − 3 = ___	**5.** 5 + 6 = ___
6. 10 − 1 = ___	**7.** 11 − 4 = ___	**8.** 2 + 7 = ___
9. 3 + 3 = ___	**10.** 8 − 7 = ___	**11.** 4 + 5 = ___
12. 10 − 3 = ___	**13.** 8 − 6 = ___	**14.** 7 − 4 = ___

15. 8 + 1   **16.** 2 + 4   **17.** 6 − 2   **18.** 9 − 4   **19.** 6 + 6

20. Use the picture. Write a number sentence to solve.

____ − ____ = ____

Chapter 1 Review/Test   thirteen **13**

Spiral Review and Test Prep
Chapter 1

Choose the best answer.

1 Skip-count by twos. What is the next number?

2, 4, 6, 8, 10, _____

○ 2   ○ 10   ○ 12   ○ 20

2 Skip-count by fives. Which number is missing?

5, 10, 15, _____, 25, 30

○ 5   ○ 10   ○ 15   ○ 20

3 Start at 4. Skip-count by 2. What is the next number? _____

Write the answer.

4 Find the difference. $9 - 6 =$ _____

5 There are 4 ducks in a pond.
3 swim away.

How many ducks are left? _____ duck

6 Write an addition or subtraction problem about this picture.

_____

_____

LOG ON www.mmhmath.com
For more Review and Test Prep

UNIT 1 CHAPTER 2
Addition Strategies and Facts to 20

Class Trip

Our class likes to go to the playground,

Our class likes to go to the zoo,

Our class likes to go to the library

And the science museum, too.

The science museum's exciting,

With so many new things to see.

Nine of my friends like the dinosaurs,

Six others the whales of the sea.

Math at Home

Dear Family,

I will learn strategies to add facts to 20 in Chapter 2. Here are my math words and an activity that we can do together.

Love, _____

My Math Words

number line :

A number line shows numbers in order.

count on :

Add 6 + 3.
Start with the greater number.
Count on 3.

addend :

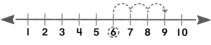

addend   addend

Home Activity

Place 7 items on the table for your child to count.

Show 1 more item. Ask how many are 7 and 1 more.

Show another item. Ask how many are 7 and 2 more.

Repeat the activity with different numbers of items.

Books to Read

Look for these books at your local library and use them to help your child learn addition facts to 20.

- **Domino Addition** by Lynette Long, Ph.D., Charlesbridge, 1996.
- **Ten Friends** by Bruce Goldstone, Henry Holt and Company, 2001.
- **I Spy Two Eyes** by Lucy Micklethwait, Morrow, William and Company, 1998.

www.mmhmath.com
For Real World Math Activities

Name _____

Order Property and Zero Property

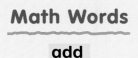

Learn Add in any order. The sum does not change when you turn around the addends.

8 is the sum of 5 + 3.

Math Words
add
sum
addend

$5 + 3 = \underline{8}$

$3 + 5 = \underline{8}$

Try It Turn around the addends. Find each sum.

$\underline{6} + \underline{3} = \underline{9}$

$\underline{3} + \underline{6} = \underline{9}$

②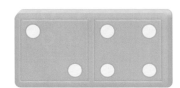

___ + ___ = ___

___ + ___ = ___

 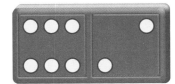

___ + ___ = ___

___ + ___ = ___

 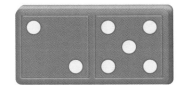

___ + ___ = ___

___ + ___ = ___

⑤ **Write About It!** Do 5 + 4 and 4 + 5 have the same sum? Why or why not?

Chapter 2 Lesson 1

Practice Find each sum.
You can add in any order.

Add 0 to a number. The sum is the same as the other addend.

6.

0 + 6 = 6

6 + 0 = 6

7.

6   0
+0   +6
─── ───
 6   6

8. 8   2
   +2   +8

9. 0   3
   +3   +0

10. 9   0
   +0   +9

11. 3   8
   +8   +3

12. 2   9
   +9   +2

13. 1   7
   +7   +1

Problem Solving **Mental Math**

14. Write the missing numbers.

4 + 3 = ☐   2 + 0 = ☐

3 + ☐ = 7   0 + ☐ = 2

Name_____   **Addition Patterns**   ALGEBRA

Learn The rule describes a pattern.
Follow the rule. Complete the table.

Math Word
rule

Rule: Add 2

In	Out
0	2
1	3
2	4
3	5

Think:

| 0 + 2 = 2 |
| 1 + 2 = 3 |
| 2 + 2 = 4 |
| 3 + 2 = 5 |

Try It Follow the rule. Add.

1. **Rule: Add 0**

In	Out
1	1
3	3

Think:

| 1 + 0 |
| 3 + 0 |

2. **Rule: Add 3**

In	Out
2	
3	

Think:

| 2 + 3 |
| 3 + 3 |

3. **Rule: Add 1**

In	Out
5	
6	
7	

Think:

| 5 + 1 |
| 6 + 1 |
| 7 + 1 |

4. **Rule: Add 4**

In	Out
0	
1	
2	

Think:

| 0 + 4 |
| 1 + 4 |
| 2 + 4 |

5. **Write About It!** Look at exercise 4. What pattern do you see?

Chapter 2 Lesson 3   twenty-one **21**

Practice — Follow the rule. Complete each table.

6 Rule: Add 4

In	Out
2	6
4	8
6	10
8	12

Think:

$2 + 4 = 6$

$4 + 4 = 8$

$6 + 4 = 10$

$8 + 4 = 12$

The rule is Add 4. Add 4 to each In number to find the Out number.

7 Rule: Add 5

In	Out
6	
7	
8	
9	

8 Rule: Add 0

In	Out
8	
9	
10	
11	

9 Rule: Add 10

In	Out
4	
6	
8	
10	

Problem Solving — Critical Thinking

10 Look at the table. What is the rule? Tell how you know.

_____

_____

In	Out
1	3
2	4
3	5
5	7

 Math at Home: Your child learned about addition patterns.
Activity: Have your child choose an exercise and tell you about the pattern.

Name_____

Game Zone

Practice at School ★ Practice at Home

Facts Path

▶ Take turns.
▶ Put your counter on start.
▶ Toss a . Move that many spaces.
▶ Find the sum.
▶ Your partner checks your answer.
▶ Move back 2 spaces if your answer is wrong.

2 players

You Will Need

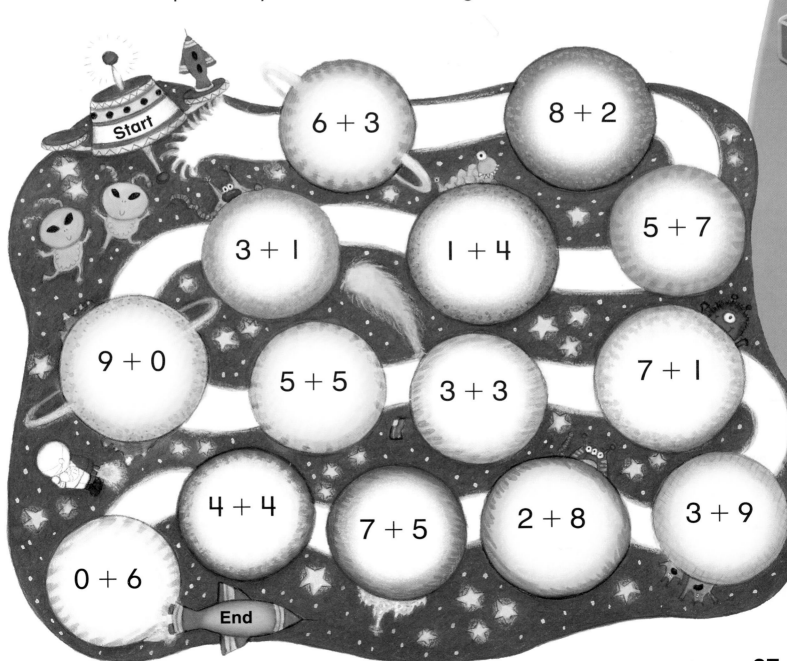

Chapter 2   Game Zone

twenty-seven **27**

Technology Link

Counting On • Calculator

You will use a  to count on.

Press.

     7

Rule: Count on 2.

Count on 2 to add.

Press.

    9

   11

   13

Count on by other numbers. Use these rules.

1. **Rule: Count on 3.**

   Press.

      8

      11

   + 3 =

   + 3 =

2. **Rule: Count on 4.**

   Press.

   + 4 =

Chapter 2
Review/Test

Name_____

Add.

 1. 6
   +2
___

2. 9
   +1
___

 3. 5
   +3
___

4. 6
    9
   +4
___

5. 3
    4
   +3
___

Turn around the addends. Find each sum.

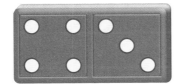

6. ___ + ___ = ___

___ + ___ = ___

7. ___ + ___ = ___

___ + ___ = ___

Complete each table.

8. **Rule: Add 2**

In	Out
0	
3	
5	

9. **Rule: Add 3**

In	Out
2	
5	
6	

Draw a picture to solve.

 10. Tom saw 5 birds on Monday. Tina saw 2 more birds than Tom on Tuesday. How many birds did they see in all? ___ birds

Spiral Review and Test Prep
Chapters 1–2

Choose the best answer.

1) What has the same sum as 8 + 4?

8 + 1	4 + 8	4 + 4	2 + 4 + 8
○	○	○	○

2) Subtract 7 − 5.

1	2	3	4
○	○	○	○

3) Count by ones.
What are the next three numbers?

6, 7, 8, 9, 10, ____, ____, ____

1, 2, 3	10, 11, 12	11, 12, 13	12, 13, 14
○	○	○	○

4) Find the missing number.
The missing number is
_____ .

Rule: Add 4

In	Out
1	5
3	7
5	?

5) June saw 9 birds.
Sara saw 3 more than June.
How many birds did Sara see? _____ birds

Tell how you found out. _____

www.mmhmath.com
For more Review and Test Prep

Subtraction Strategies and Facts to 20

UNIT 1 CHAPTER 3

Twelve Little Rabbits

Twelve little rabbits in a rabbit pen;

Two hopped away, and then there were ten.

Ten little rabbits with ears up straight;

Two hopped away, and then there were eight.

Eight little rabbits doing funny tricks;

Two hopped away, and then there were six.

Two little rabbits found a new friend;

They hopped away, and that is the end.

Math at Home

Dear Family,

I will learn strategies for subtracting facts to 20 in Chapter 3. Here are my math words and an activity that we can do together.

Love, _____

My Math Words

difference:
6 − 1 = 5

count back:
Subtract 6 − 1.
Start with 6 and count back 1.

6 − 1 = 5

related facts:
5 + 1 = 6
6 − 1 = 5

Home Activity

Place up to 6 items on the table for your child to count.

Take away one item. Ask how many there are now.

Take away two more items. Ask how many are left.

Books to Read

Look for these books at your local library and use them to help your child subtract to 20.

- **3 Pandas Planting** by Megan Halsey, Bradbury Press, 1994.
- **Safari Park** by Stuart J. Murphy, HarperCollins, 2002.
- **Twenty Is Too Many** by Kate Duke, Dutton Children's Books, 2000.

www.mmhmath.com
For Real World Math Activities

Name_____

Count Back to Subtract

Learn Use a number line to count back to subtract. Subtract to find the difference.

Math Words
count back
subtract
difference

Count back 3. Start at 12.

12 − 3 = __9__

12 − 3 = 9 and 9 = 12 − 3 are the same.

__9__ = 12 − 3

Try It Count back to subtract. You can use the number line.

① 11 − 3 = __8__ ② 10 − 1 = ___ ③ 6 − 2 = ___

④ 8 − 2 = ___ ⑤ 11 − 2 = ___ ⑥ 10 − 3 = ___

⑦ ___ = 9 − 2 ⑧ ___ = 7 − 1 ⑨ ___ = 8 − 1

⑩ ✏️ **Write About It!** How can you count back to subtract?

Chapter 3 Lesson 1 thirty-three **33**

Practice Count back to subtract. You can use the number line.

Count back 1, 2, or 3 to subtract.

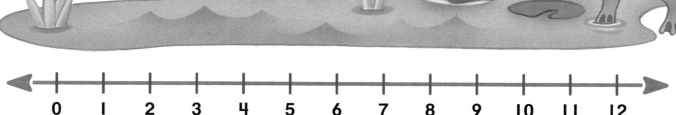

11) 7 − 2 = 5   12) 10 − 3 = ___   13) 8 − 3 = ___

14) 9 − 1 = ___   15) 5 − 1 = ___   16) 12 − 3 = ___

17) 9 − 2 = ___   18) 10 − 2 = ___   19) 7 − 3 = ___

20) ___ = 8 − 2   21) ___ = 7 − 0   22) ___ = 9 − 3

23) ___ = 10 − 3   24) ___ = 4 − 2   25) ___ = 9 − 1

Problem Solving **Mental Math**

26) 6 birds in a tree. 3 birds fly away. How many birds are left?

___ birds

27) 12 frogs and 3 ducks are in a pond. How many more frogs than ducks are in the pond?

___ frogs

Math at Home: Your child used a number line to subtract 1, 2, or 3 from a number.
Activity: Say a number between 3 and 12. Have your child subtract 1, 2, or 3. Have your child count back to find the difference.

Name_____

Subtract All and Subtract Zero

Learn You can subtract a number from itself.
You can subtract 0 from a number.

6 – 6 = __0__

6 – 0 = __6__

The difference is 0.

The difference is the same as the number.

Try It Subtract.

1. 8 – 8 = __0__

   8 – 0 = __8__

2. 7 – 7 = ____

   7 – 0 = ____

3. 9 – 9 = ____

   9 – 0 = ____

4. 5 – 5 = ____

   5 – 0 = ____

5. 10 – 10 = ____

   10 – 0 = ____

6. 12 – 12 = ____

   12 – 0 = ____

7. **Write About It!** Subtract 4 from 4. What is the answer? Draw a picture and explain your answer.

Practice Subtract. If the difference is 0, color 🖍.
Color the other parts any color you like.

$$\begin{array}{r}6\\-6\\\hline\end{array}$$

$$\begin{array}{r}8\\-1\\\hline 7\end{array}$$
$$\begin{array}{r}12\\-3\\\hline\end{array}$$
$$\begin{array}{r}6\\-0\\\hline\end{array}$$
$$\begin{array}{r}7\\-2\\\hline\end{array}$$

$$\begin{array}{r}9\\-9\\\hline\end{array}$$
$$\begin{array}{r}11\\-2\\\hline\end{array}$$
$$\begin{array}{r}7\\-7\\\hline\end{array}$$
$$\begin{array}{r}9\\-3\\\hline\end{array}$$
$$\begin{array}{r}10\\-10\\\hline\end{array}$$

$$\begin{array}{r}7\\-0\\\hline\end{array}$$
$$\begin{array}{r}8\\-2\\\hline\end{array}$$
$$\begin{array}{r}10\\-0\\\hline\end{array}$$
$$\begin{array}{r}20\\-0\\\hline\end{array}$$
$$\begin{array}{r}20\\-20\\\hline\end{array}$$

Make it Right

8 Bob wrote this number sentence.

$5 - 0 = 0$

Why is Bob wrong? Make it right.

Math at Home: Your child subtracted 0 from numbers to 20 and also subtracted numbers from themselves.
Activity: Ask your child to use small items to show $4 - 4 = 0$ and $4 - 0 = 4$.

Name_____

Relate Addition to Subtraction

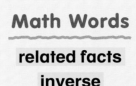

Learn You can use addition facts to subtract.
Related facts have the same three numbers.

Math Words
related facts
inverse

4 + 5 = 9

Addition and subtraction are inverse operations.

9 − 5 = 4

Your Turn Use 🖍 and 🖍 to show the addition fact.
Find the sum and difference for the facts.

1. 6 + 2 = 8   8 − 2 = 6

2. 2 + 9 = ___   11 − 9 = ___

3. 4 + 4 = ___   8 − 4 = ___

4. ✏️ **Write About It!** Write a related subtraction fact for 5 + 7 = 12.

Chapter 3 Lesson 3   thirty-seven **37**

Practice Add and subtract. Circle the related facts. You can use .

You can use addition facts to subtract.

5. 7   12
   +5   − 7
   ─── ───
   12   5   (circled as related facts)

6. 8   10
   +2   − 2

7. 9   9
   +3   −3

8. 6   11
   +5   − 5

9. 9   14
   +5   − 5

10. 8   15
    +7   − 7

11. 3   11
    +8   − 3

12. 9   9
    +2   −2

13. 5   9
    +4   −4

14. 6   10
    +4   − 6

15. 10   20
   +10 −10

16. 7   14
    +7   − 7

Spiral Review and Test Prep

17. There are 5 birds inside the birdhouse. 2 more birds fly in. How many birds in all?

    4   6   7   8
    ○   ○   ○   ○

18. There are 2 butterflies. Each butterfly has 8 spots. How many spots in all?

    8   10   16   20
    ○   ○   ○   ○

Math at Home: Your child learned how to use addition facts to subtract.
Activity: Put 12 objects on the table. Cover some of them. Have your child use subtraction to say how many are missing. For example, 12 − 3 = 9.

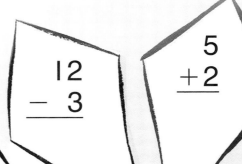

Name_____

Find each sum or difference. Color.
Answers greater than 10: 🖍▶
Answers less than 10: 🖍▶
Answers equal to 10: 🖍▶

12
− 3

5
+2

6
+0

9
+2

8
−8

9
+1

15
− 7

8
+2

7
+1

3
+8

7
+2

2
+8

13
− 9

15
− 8

10
− 0

9
+4

8
+0

9
−7

10
− 3

12
− 2

Chapter 3 Extra Practice

thirty-nine **39**

Extra Practice

Count on to add.
9 + 3 = 12

Count back to subtract.
12 − 3 = 9

Use related facts.
6 + 8 = 14
14 − 8 = 6

Add in any order.
7 + 2 = 9
2 + 7 = 9

1. 1 + 9 = ___
2. 7 + 3 = ___
3. 12 − 9 = ___
4. 9 − 9 = ___
5. 2 + 8 = ___
6. 10 − 3 = ___
7. 8 − 0 = ___
8. 9 − 3 = ___
9. 7 + 6 = ___

10. 8
    +1

11. 11
    − 2

12. 12
    − 8

13. 2
    +7

14. 9
    −7

15. 2
    +9

16. 11
    − 9

17. 8
    +8

18. 13
    − 8

19. 10
    −10

20. **Write About It!** How would you find the difference from 11 − 2? Use pictures or words to explain.

www.mmhmath.com
For more Practice

Math at Home: Your child practiced addition and subtraction facts.
Activity: Copy ten of the facts. Ask your child to write the answers.

Name_____ **Missing Number**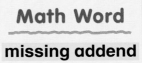

 Find the missing addend.

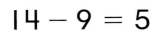 ? 9 + ? = 14

Use a related subtraction fact to help you find the missing addend.

Math Word
missing addend

14 − 9 = 5

So 9 + 5 = 14.

The missing addend is 5.

Try It Use related facts. Find the missing addend. You can draw dots to help.

 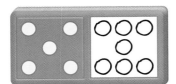

12 − 5 = __7__

5 + 7 = 12

17 − 8 = ____

8 + ☐ = 17

13 − 6 = ____

6 + ☐ = 13

15 − 9 = ____

9 + ☐ = 15

 Write About It! How do you find the missing addend in 8 + ☐ = 13? Use pictures or words to explain.

Practice Find the missing number.

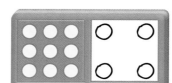

6) $13 - 9 = \underline{4}$

$9 + \boxed{4} = 13$

> $13 - 9 = 4$ and $9 + 4 = 13$ are related facts.

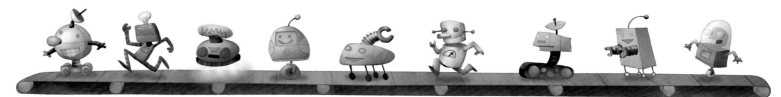

7) $15 - 7 = \boxed{8}$

$7 + \boxed{8} = 15$

8) $13 - 8 = \boxed{}$

$8 + \boxed{} = 13$

9) $9 + 7 = \boxed{}$

$16 - \boxed{} = 7$

10) $3 + 4 = \boxed{}$

$7 - \boxed{} = 3$

11)
```
    6              12
  + ☐            - ☐
  ----          ----
   12              6
```

12)
```
    8              14
  + ☐            - ☐
  ----          ----
   14              6
```

Problem Solving — Mental Math

Find the missing number.

13) $\boxed{} + 4 = 8$

14) $8 + \boxed{} = 16$

15) $10 + \boxed{} = 20$

16) $\boxed{} + 9 = 18$

17) $7 + \boxed{} = 14$

18) $\boxed{} + 5 = 10$

Math at Home: Your child used related facts to find missing numbers.
Activity: Ask your child to tell you the missing addend that will help him or her add $9 + \boxed{} = 16$.

Chapter 3
Review/Test

Name _____

Subtract.

1. $10 - 2 =$ ____
2. $11 - 3 =$ ____
3. $12 - 1 =$ ____
4. $9 - 3 =$ ____
5. $12 - 3 =$ ____
6. $11 - 2 =$ ____

Add or subtract.

7. $10 - 8$
8. $14 - 6$
9. $10 - 10$
10. $8 - 0$
11. $12 - 3$
12. $3 + 9$

13. $2 + 8$
14. $10 - 2$
15. $11 - 2$
16. $7 - 0$
17. $7 - 7$
18. $20 - 20$

Solve.

19. 9 ducks are in the pond. 3 ducks swim away. How many ducks are in the pond now?

____ ducks

20. Joe saw 11 birds. Leo saw 3 birds. How many more birds did Joe see than Leo?

____ more birds

Spiral Review and Test Prep
Chapters 1–3

1. There are 11 flowers in all. 6 are red. The rest are pink. How many are pink?

   ○ 5   ○ 6   ○ 11   ○ 17

2. $9 + 2 =$ ___

   ○ 9   ○ 10   ○ 11   ○ 12

3. Which shows related facts?

$8 + 2 = 10$	$3 + 4 = 7$	$6 + 3 = 9$	$5 + 5 = 10$
$10 - 2 = 8$	$4 - 3 = 1$	$6 - 3 = 3$	$5 - 5 = 0$
○	○	○	○

4. What is the missing number?

   Rule: Add 5

In	Out
2	7
4	9
6	?

   The missing number is ___.

5. 4 turtles are on a log. 3 more turtles come. Then 1 turtle swims away. How many turtles are on the log now? Draw pictures to explain.

   ___ turtles

www.mmhmath.com
For more Review and Test Prep

Exploring Addition and Subtraction

Bubbles

Sung to the tune of "How Much Is That Doggie in the Window?"

Today I blew thirteen soapy bubbles.

Ten popped! Then I had only three.

The next time I try to blow some bubbles,

I won't stand so close to a tree.

Math at Home

Dear Family,

I will learn ways to help me add and subtract in Chapter 4. Here are my math words and an activity we can do together.

Love, _____

My Math Words

doubles:

$8 + 8 = 16$

make a ten:

$9 + 3 = 12$

10 + 2 is the same as 9 + 3.

fact family:

Each fact in a fact family uses the same numbers.

$3 + 2 = 5 \quad 5 - 2 = 3$
$2 + 3 = 5 \quad 5 - 3 = 2$

Home Activity

Make number cards for 1–5. Shuffle and place the cards facedown. You and your child each choose one card. Count out objects to match the numbers. Have your child write a number sentence to find the total.

$3 + 2 = 5$

Books to Read

In addition to these library books, look for the Time For Kids math story that your child will bring home at the end of this unit.

- **How Many Feet? How Many Tails?** by Marilyn Burns, Scholastic, 1996.
- **Mission Addition** by Loreen Leedy, Holiday House, Inc., 1999.
- **Time For Kids**

LOG ON www.mmhmath.com
For Real World Math Activities

Name_____

Use Doubles to Add and Subtract

Learn You can use doubles to find a sum.

Math Words
doubles
doubles plus 1

6 + 6 = __12__
doubles

6 + 7 = __13__
doubles plus 1

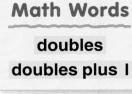
6 + 6 is 12.
6 + 7 is 1 more.
So 6 + 7 is 13.

You can use doubles to find a difference.

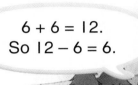
6 + 6 = 12.
So 12 − 6 = 6.

6 + 6 = __12__

12 − 6 = __6__

Try It Add or subtract.

1. 5 + 5 = ___
2. 5 + 6 = ___
3. 10 − 5 = ___

4. 7 + 7 = ___
5. 7 + 8 = ___
6. 14 − 7 = ___

7. 4 + 4 = ___
8. 4 + 5 = ___
9. 8 − 4 = ___

10. **Write About It!** What doubles fact can you use to solve 16 − 8?

Chapter 4 Lesson 1

fifty-three **53**

Practice Add or subtract.

> The addends are the same in a doubles fact.

11. 8 +8 = 16	12. 8 +9	13. 16 − 8	14. 9 +9	15. 9 +10
16. 10 +10	17. 10 + 7	18. 20 −10	19. 5 +6	20. 11 − 6
21. 7 +8	22. 2 +3	23. 10 − 5	24. 12 − 6	25. 7 +7
26. 4 +3	27. 18 − 9	28. 6 +7	29. 13 − 7	30. 6 +6

Problem Solving — Number Sense

31. Pete saw 4 birds. Toby saw the same number of birds. How many birds did they see in all?

____ + ____ = ____

____ birds

32. Joe counts 7 deer. Kay counts one more deer than Joe. How many deer do they count in all?

____ + ____ = ____

____ deer

Math at Home: Your child used doubles, such as 6 + 6 = 12, to add and subtract.
Activity: Give your child an addition doubles fact such as 9 + 9. Have your child solve the fact, and then tell a related subtraction fact.

Name_____

Use 10 to Add and Subtract 9

HANDS ON Activity

Learn You can make a ten to help you add.

Add 9 + 3.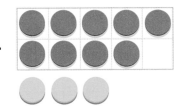

Math Word
make a ten

10 + 2 = __12__

9 + 3 = __12__

I make a ten and have 2 ones. 10 + 2 is the same as 9 + 3.

Your Turn Use ● and a ▭. Draw dots to make a ten. Add.

1. 9 + 2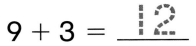

   10 + __1__

   9 + 2 = __11__

2. 9 + 5

   10 + ___

   9 + 5 = ___

3. 9 + 6

   10 + ___

   9 + 6 = ___

4. 9 + 9

   10 + ___

   9 + 9 = ___

5. ✏️ **Write About It!** Explain how making 10 helps you add 9.

Chapter 4 Lesson 2

Practice Cross out 9 to subtract.

6) 14 − 9 = 5

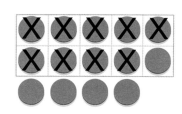

> 14 − 10 is 4.
> 10 is 1 more than 9.
> So 14 − 9 is 5.

7) 11 − 9

8) 16 − 9

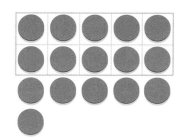

Add or subtract. You can use ● and a ▭.

9) 9 + 4

10) 12 − 9

11) 9 + 8

12) 15 − 9

13) 9 + 7

14) 13 − 9

15) 15 − 9

16) 9 + 7

17) 17 − 9

18) 19 − 9

Problem Solving **Number Sense**

19) Bill caught 9 bugs. Jean caught 3 bugs. How many bugs did they catch in all?

_____ bugs

Tell how you solved the problem.

_____

_____

Math at Home: Your child learned how to add and subtract 9 using a ten-frame and counters.
Activity: Have your child show you how to use a ten-frame to add 9 + 6 and subtract 13 − 9.

Name_____

Use 10 to Add and Subtract 7 and 8

Learn You can use a ☐☐☐☐☐ to help you add or subtract 7 and 8.

Subtract 13 − 8.

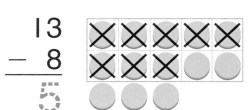

13
− 8
―――
  5

13 − 10 is 3.
10 is 2 more than 8.
So 13 − 8 is 5.

Your Turn Subtract. Use ● and a ☐☐☐☐☐.

1.
15
− 7
―――
  8

2.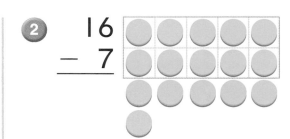
16
− 7

3.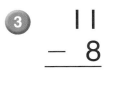
11
− 8

4.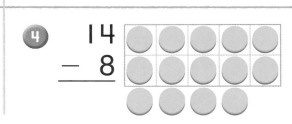
14
− 8

Draw dots to make a ten. Add.

5. 8 + 5

10 + ___

8 + 5 = ___

6. 7 + 5

10 + ___

7 + 5 = ___

7. **Write About It!** How many counters do you move into the ten-frame when you add 7 + 4? Explain.

Chapter 4 Lesson 3   fifty-seven **57**

Practice Add or subtract.
You can use ⬤ and a ▦.
Color answers 8 or more 🖍▶.
Color answers 7 or less 🖍▶.

8 + 2

7 + 9

11 − 8

15 − 7

13 − 8

17 − 8

7 + 6

8 + 7

11 − 7

8 + 4

13 − 7

15 − 8

Math at Home: Your child learned how to add and subtract 7 and 8 using a ten-frame and counters.
Activity: Have your child show you how to use a ten-frame to add 8 + 9 and subtract 13 − 7.

Extra Practice

Name_____

Complete the addition table. Circle the sums for the doubles facts. Draw a square around the doubles plus one fact.

+	0	1	2	3	4	5	6	7	8	9
0	(0)	[1]								
1										
2	2	3	4							
3									11	12
4		5				9	10			13
5							11			
6										
7				10	11	12			15	
8						13				17
9										18

Chapter 4 Extra Practice

Extra Practice

Look for a pattern.
Write the missing numbers.

2, 4, 6, 8, 10

3, 6, 9, ___, ___

5, 10, 15, ___, ___

14, 16, 18, ___, ___

www.mmhmath.com
For more Practice

Math at Home: Your child practiced addition facts and skip-counting.
Activity: Write out addition problems from the addition table on page 61 and ask your child to solve them.

Name_____

Game Zone

Practice at School ★ Practice at Home

On a Nature Walk

▶ You and a partner take turns.

▶ Put your ● on **Start**.

▶ Toss a . Move that many spaces.

▶ Find the sum or difference.
Your partner uses a calculator to check the answer.

▶ If correct, stay in the space.
If not, move back 2 spaces.

▶ Play until you reach the **Waterfall**.

2 players

You Will Need
1 🎲
2 ●
🖩

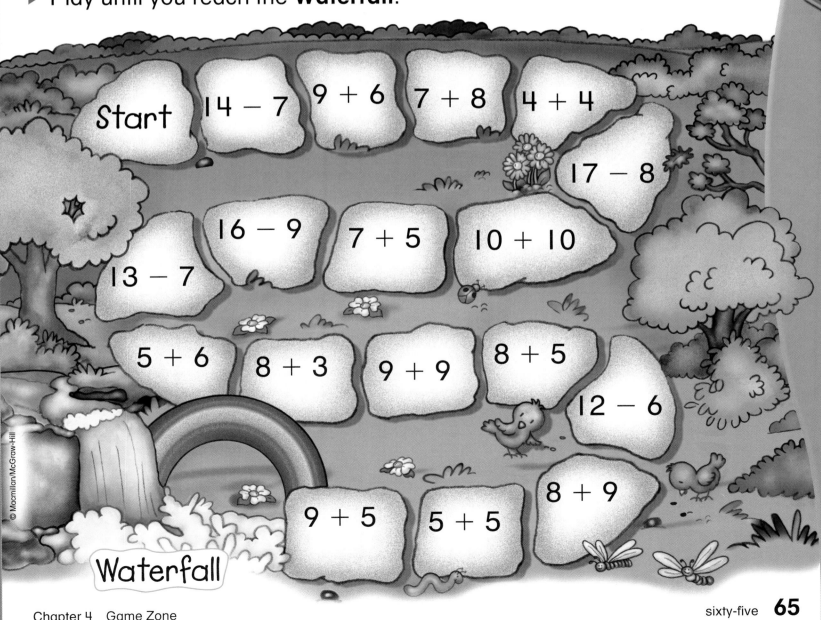

Chapter 4 Game Zone

sixty-five **65**

Technology Link

Model Addition and Subtraction • Computer

Use [stamp] to show ways to make 11.

- Choose a mat to add.
- Stamp out 6 butterflies.
- Stamp out 5 butterflies.
- Click on +.

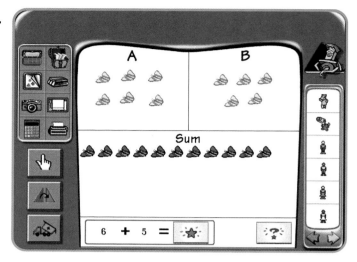

What addition fact do you see? $\underline{6} + \underline{5} = \underline{11}$

- Next, stamp out 7 butterflies and 4 butterflies.

  What addition fact do you see?   ___ + ___ = ___

- Stamp out 8 butterflies and 3 butterflies.

  What addition fact do you see?   ___ + ___ = ___

How are the facts alike? _____

Subtract. You can use the computer.
Choose a mat to subtract.

| 13 – 4 = ___ | 10 – 5 = ___ |
| 12 – 6 = ___ | 17 – 8 = ___ |

 For more practice use Math Traveler.™

Name _____

TIME FOR KIDS

17 sea horses swim together.
Then 9 hide inside the treasure chest.

$17 - 9 = 8$

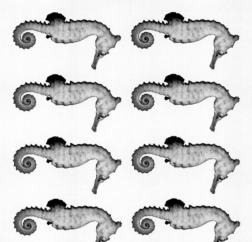

The bowl has 10 fish.
Write two facts about the fish.

$5 + 5 = 10$

$10 - 5 = 5$

Fold down
- -

TIME FOR KIDS

Sea Facts

Joke: Why are fish so smart?

Answer: They live in schools!

READ TOGETHER

© Macmillan/McGraw-Hill

A group of fish is called a school.
You can see schools of fish at an aquarium.

12 fish are swimming.
One school is yellow.
The other is orange and white.

6 + 6 = 12

3 dolphins are at the top of the tank.
4 other dolphins swim to the bottom of the tank.

3 + 4 = 7

Three schools of fish are swimming below.
In all there are 11 fish.

3 + 6 + 2 = 11

Name_____

Unit 1 Performance Assessment

Number Story

Tyler is juggling 5 balls. Amy is juggling 4 balls.

Finish this story with an addition or subtraction problem. Draw a picture or write the problem. Write the number sentence that solves the problem.

Show your work.

 You may want to put this page in your portfolio.

Unit 1 Performance Assessment

Unit 1
Enrichment

Missing Numbers

Which number could go in the box?

5 + ☐ > 10

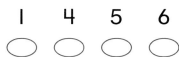
1   4   5   6
○   ○   ○   ○

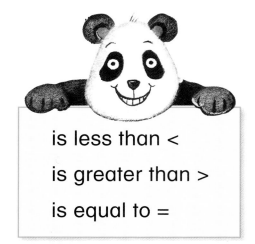

Which number added to 5 is greater than 10?

is less than <
is greater than >
is equal to =

5 + 1 = 6,   5 + 4 = 9,   5 + 5 = 10,   5 + 6 = 11

So 6 is the correct answer.

Choose the best answer.

1 Which number could go in the box?

2 + ☐ > 10

2   5   8   9
○   ○   ○   ○

---

2 Which number could go in the box?

4 + ☐ < 10

5   7   8   9
○   ○   ○   ○

Show your work.

UNIT 2 CHAPTER 5

Understanding Place Value

Snap and Tap!

Story by Marsha Comito
Illustrated by John Nez

Snap and tap! Ready to go.
10 of us march in a row.

Circle a group of 10 children.

Snap and tap! Stay in line.
All the drums sound just fine.

Circle groups of 10 drums.

Snap and tap! Tap your toes.
All the horns are set to blow.

Circle groups of 10 horns.

Snap and tap, fast and slow.
Lots of tambourines in a row.

Circle groups of 10 tambourines.

Math at Home

Dear Family,

I will learn about tens and ones and numbers to 100 in Chapter 5. Here are my math words and an activity that we can do together.

Love, _____

My Math Words

digit :
56
5 and 6 are the digits.

tens and ones :
56
↓ 6 ones
5 tens

place value :
56
↓ 6
50

Home Activity

Ask your child to take a handful of beans and place them on a table. Have your child count the number of beans and tell how many there are.

Repeat the activity a few times.

Look for these books at your local library and use them to help your child learn numbers to 100.

- **100th Day Worries** by Margery Cuyler, Simon & Schuster, 2000.
- **The Wolf's Chicken Stew** by Keido Kasza, Putnam Publishing Group, 1996.
- **100 School Days** by Anne Rockwell, HarperCollins, 2002.

www.mmhmath.com
For Real World Math Activities

Name _____ Tens

Learn You can group ones to make tens.

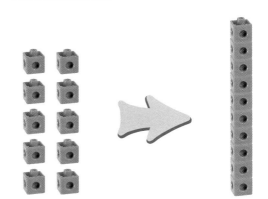

Another name for 10 ones is 1 ten.

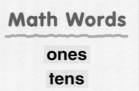

Math Words
ones
tens

10 ones = __1__ ten

Your Turn Use to make groups of ten. Write how many tens.

1. 20 ones = __2__ tens

2. 30 ones = ____ tens

3. 40 ones = ____ tens

4. 50 ones = ____ tens

5. 60 ones = ____ tens

6. 70 ones = ____ tens

7. **Write About It!** How many tens are there in 80? How do you know?

Chapter 5 Lesson 1 seventy-seven 77

Practice Write how many tens and how many ones. Then write the number.

 You can trade 1 ten for 10 ones.

⑧ 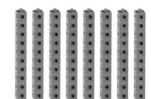 __8__ tens = __80__ ones = __80__

⑨ ____ tens = ____ ones = ____

⑩ ____ tens = ____ ones = ____

⑪ ____ tens = ____ ones = ____

⑫ ____ tens = ____ ones = ____

Problem Solving **Number Sense**

⑬ Mike trades 10 ones for 1 ten. Then he trades 10 tens for 1 hundred.

Write the number. _____

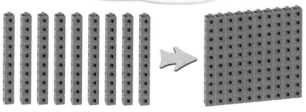 10 groups of ten is 100.

Name_____ **Tens and Ones**

Learn You can trade 23 ones for 2 tens 3 ones.

23 ones = __2__ tens __3__ ones

tens	ones
2	3

Your Turn Write how many tens and ones.

1. 17 ones = __1__ ten __7__ ones

tens	ones
1	7

2. 25 ones = ____ tens ____ ones

tens	ones

3. 36 ones = ____ tens ____ ones

tens	ones

4. **Write About It!** How is 42 different from 24? Draw pictures or use words to explain.

Chapter 5 Lesson 2   seventy-nine **79**

Practice Each picture shows 2 tens 3 ones.

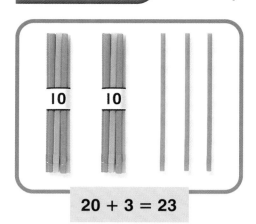
20 + 3 = 23

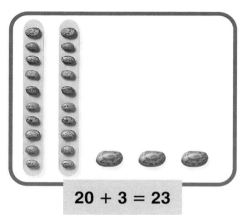
20 + 3 = 23

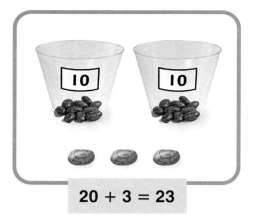
20 + 3 = 23

Write how many tens and ones. Then write the number.

5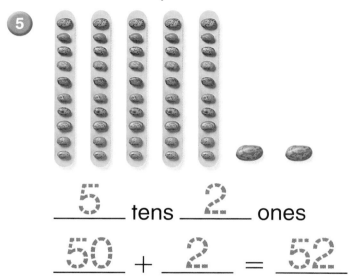

___5___ tens ___2___ ones

___50___ + ___2___ = ___52___

6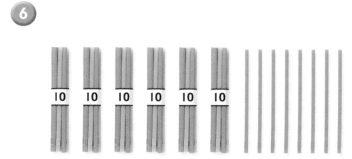

_____ tens _____ ones

_____ + _____ = _____

7

_____ tens _____ one

_____ + _____ = _____

8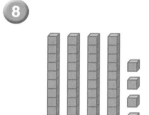

_____ tens _____ ones

_____ + _____ = _____

9 40 + 9 = _____

10 70 + 3 = _____

Math at Home: Your child learned about tens and ones in 2-digit numbers.
Activity: Write some 2-digit numbers, like 53, and have your child tell you how many tens and how many ones (5 tens 3 ones).

Name_____

Place Value to 100

Learn You can show the place value of 28. The place of a digit in a number tells its value.

Math Words
place value
digit

Try It Circle the value of the green digit.

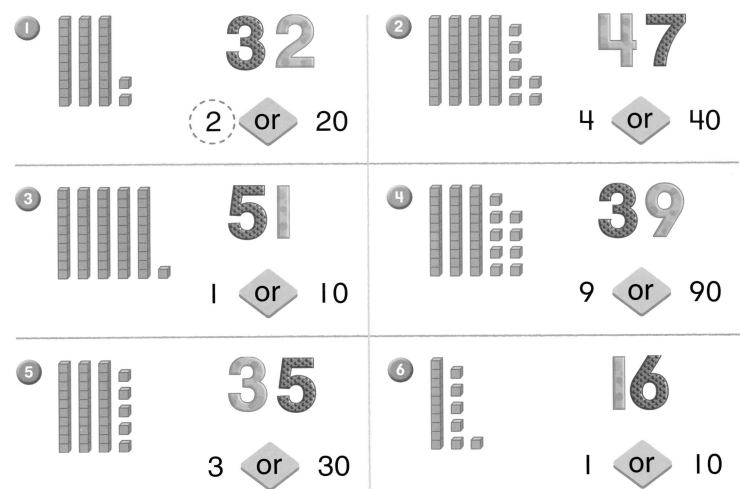

1. 32 — (2) or 20
2. 47 — 4 or 40
3. 51 — 1 or 10
4. 39 — 9 or 90
5. 35 — 3 or 30
6. 16 — 1 or 10

7. **Write About It!** What is the value of the digit 7 in the number 27? Explain.

Chapter 5 Lesson 3   eighty-one **81**

Practice Circle the value of the green digit.

8.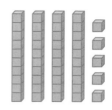
(5) or 50

9.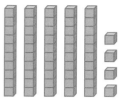
5 or 50

10. 68
6 or 60

11. 24
4 or 40

12. 97
9 or 90

13. 37
7 or 70

14. 12
1 or 10

15. 36
3 or 30

16. 84
8 or 80

17. 42
2 or 20

18. 59
9 or 90

Problem Solving **Number Sense**

Draw lines to match the same numbers.

19. 40 + 2

20. 20 + 2

21. 1 ten 5 ones

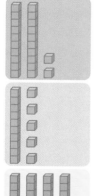

Math at Home: Your child learned the value of each digit in 2-digit numbers.
Activity: Say some 2-digit numbers, for example, 28. Have your child tell the place value of each digit (2 = 2 tens; 8 = 8 ones).

Name_____   **Read and Write Numbers**

Learn You can write numbers as words.
You can write these words as numbers too.

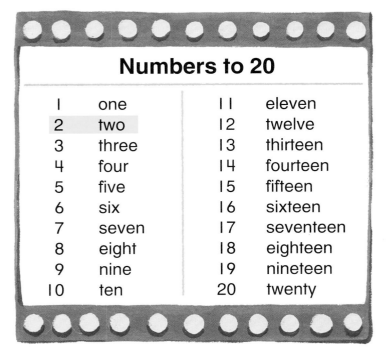
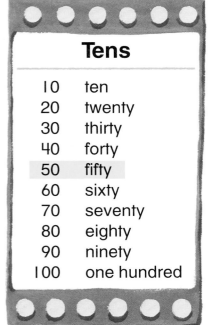

Fifty-two is 52.

Try It Write each word as a number.

1. seven __7__

2. fourteen _____

3. ninety _____

4. thirty _____

5. seventy-five _____

6. sixty-two _____

7. eighty-three _____

8. twenty-four _____

9. forty-nine _____

10. fifty-seven _____

11.  **Write About It!** How do you write 45 as a word?

Chapter 5 Lesson 4

eighty-three **83**

Practice Write each as a number word.

Look back at the chart if you need help.

12. 21 _twenty-one_

13. 80 _____
14. 11 _____
15. 52 _____
16. 9 _____
17. 15 _____
18. 46 _____
19. 70 _____
20. 39 _____
21. 87 _____

Spiral Review and Test Prep

22. How many straws in all?

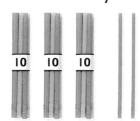

23 ○   30 ○   32 ○

23. 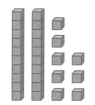 28 is the same as

20 + 8 ○   2 tens ○   8 tens 2 ones ○

 Math at Home: Your child learned to read and write numbers to 100 as words.
Activity: Write some 2-digit numbers, like 47. Have your child read the number and write the number word.

Name _____

Estimate Numbers

 About how many marbles will fill the large jar? Estimate to find out.

Math Word
estimate

Use the small jar to help you estimate. 10 marbles are in the small jar.

About 4 small jars will fill a large jar. 10, 20, 30, 40

About 40 marbles will fill the large jar.

Try It About how many marbles will fill the large jar?

1.

   about __20__ marbles

2.

   about _____ marbles

3. ✏️ **Write About It!** How did you estimate?

Chapter 5 Lesson 5

Practice About how many are in each picture? Circle your estimate.

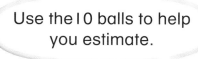

Use the 10 balls to help you estimate.

4.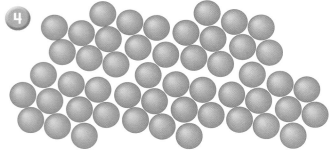

about 10   (about 50)

5.

about 20   about 50

6.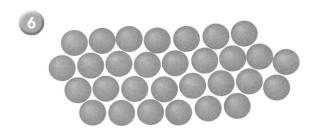

about 30   about 90

7.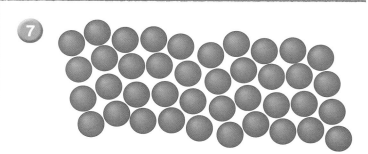

about 10   about 40

Problem Solving — Estimation

Circle the best answer.

8. Ron has a pile of marbles. He estimates that he has about 20 marbles. Is Ron's estimate reasonable? Explain.

_____   Yes   No

Math at Home: Your child estimated how many objects are in a group.
Activity: Put 10 beans in a group. Then have your child put a handful of beans in another group and estimate how many beans there are.

86 eighty-six

Name_____

Problem Solving Skill
Reading for Math

The School Band

60 people came to the concert. Then 20 more people came late. There are 7 rows of chairs. Each row has 10 chairs.

Reading Skill **Make Predictions**

1. Do you think everyone who came late found a chair? Explain.

   _____

   _____

2. How many people come to the concert in all?

   _____ people

3. How many more rows of chairs are needed?

   _____ more row

The Concert

The band has 12 children. Next year, 3 more children will join the band. There are 17 hats.

Problem Solving

Reading Skill — **Make Predictions**

4) How many more children will be in the band next year?

_____ children

Write the addition sentence.

_____ + _____ = _____

5) The school has 17 band hats.
Write the number of hats as tens and ones.

_____ tens _____ ones

6) Do you think the school will have to buy more hats next year?

_____

_____

Math at Home: Your child made predictions to answer questions.
Activity: Have your child predict something that will happen tomorrow and explain why he or she thinks it will happen.

Problem Solving Practice

Name _____

Solve.

1 8 🛼 are in the box.
Tom puts 4 more in the box.
How many 🛼 are in the box?

8 + 4 = _____ 🛼

Write the total number of 🛼 as tens and ones.

_____ ten _____ ones

2 7 friends went ice skating.
6 more friends joined them.
How many skaters were there?

_____ + _____ = _____

_____ skaters

Write the total number of skaters as tens and ones.

_____ ten _____ ones

Write a Story!

3 Write how many tens and ones.
Then write a story about 24 things!

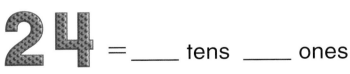 = _____ tens _____ ones

_____

_____

_____

Chapter 5 Problem Solving Practice

Writing for Math

Write a story about the number of drums in the picture.

Use the words on the cards to help.

row   ten

Think

How many rows of ten drums do I see in this picture? _____

How many others? _____

How many drums are there all together? _____

_____ tens _____ ones = _____

Solve

I can write my number story now.

_____

_____

_____

Explain

I can tell you how my number matches my story.

_____

_____

_____

Chapter 5
Review/Test

Name_____

Write how many tens and ones.
Then write the number.

① ___ tens ___ ones

___ + ___

___

② ___ tens ___ ones

___ + ___

___

Circle the value of the green digit.

③ 12   1   or   10

④ 38   3   or   30

⑤ 63   3   or   30

About how many balls are in each picture?

⑥ about 10   about 50

⑦ about 10   about 30

Write each number.

⑧ fifty-nine _____

⑨ thirty-one _____

⑩ sixteen _____

Chapter 5   Review/Test

ninety-one **91**

Spiral Review and Test Prep
Chapters 1–5

Choose the best answer.

1. What is the same as 8 + 4?

8 + 1	4 + 8	4 + 4	2 + 4 + 8
○	○	○	○

2. There are 8 chicks inside the barn. How many chicks in all?

4	6	8	12
○	○	○	○

3. Write the number word for 15. _____

4. Look at the table.

What is the rule? _____.

Input	Output
0	4
1	5
3	7
5	9

5. 9 flowers float on lily pads.
5 more flowers bloom.
How many flowers are there?

_____ flowers
Tell how you know.

_____

www.mmhmath.com
For more Review and Test Prep

UNIT 2 CHAPTER 6
Numbers and Patterns

Skip-Count Song

Sung to the tune of "Skip to My Lou"

Skip-count, skip-count, count by 10s
Skip-count, skip-count, count by 10s
Skip-count, skip-count, count by 10s
We can count to 100.
10, 20, 30, 40, 50, 60, 70, 80, 90, 100!

Math at Home

Dear Family,

I will learn about comparing and ordering numbers to 100 in Chapter 6. Here are my math words and an activity that we can do together.

Love, _____

My Math Words

is less than (<) :
2 < 5

is greater than (>) :
5 > 2

ordinal numbers :
The cat is first in line.

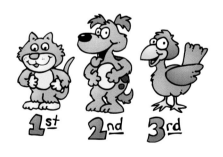

Home Activity

Place 32 small items on the table. Have your child make groups of ten. Ask how many tens there are; how many ones?

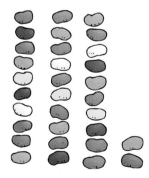

Repeat with different numbers.

Books to Read

Look for these books at your local library and use them to help your child learn numbers and patterns.

- **Count on Pablo** by Barbara deRubertis, Kane Press, 1999.
- **The Biggest Fish** by Sheila Keenan, Scholastic Inc, 1996.
- **Missing Mittens** by Stuart J. Murphy, HarperCollins, 2001.

www.mmhmath.com
For Real World Math Activities

Name_____   **Compare Numbers**

Learn You can compare numbers.

Compare the tens first. If the tens are the same, then compare the ones.

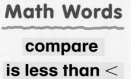

Math Words
compare
is less than <
is greater than >
is equal to =

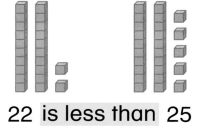

22 is less than 25

22 < 25

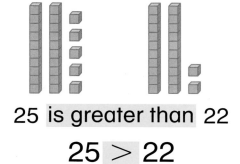

25 is greater than 22

25 > 22

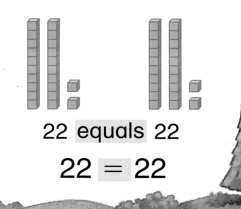

22 equals 22

22 = 22

Try It Compare. You can use  and ▫.

①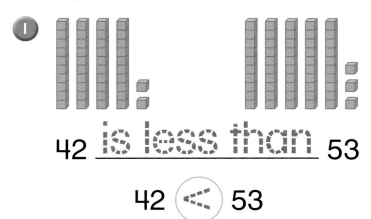

42 ___is less than___ 53

42 < 53

②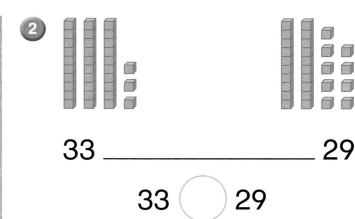

33 _____ 29

33 ◯ 29

③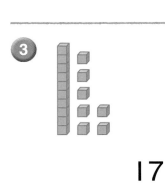

17 ◯ 17

④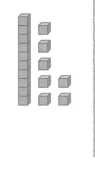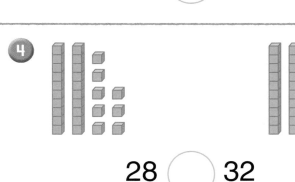

28 ◯ 32

⑤ **Write About It!** How can you tell that 36 is greater than 35?

Chapter 6 Lesson 1   ninety-five **95**

Practice Compare. You can use 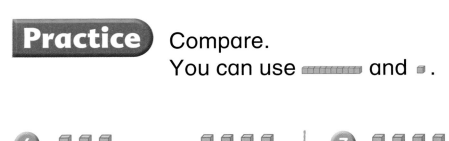 and ▪.

Remember to compare the tens first.

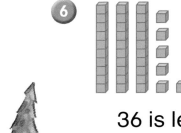
6

36 is less than 40
36 < 40

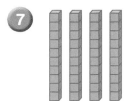 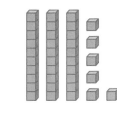
7

40 is greater than 36
40 > 36

8 76 is __greater than__ 48   76 ◯ 48

9 42 is _____ 51   42 ◯ 51

10 40 ◯ 40   11 79 ◯ 71   12 36 ◯ 51

13 26 ◯ 22   14 95 ◯ 99   15 49 ◯ 49

16 73 ◯ 8   17 67 ◯ 67   18 46 ◯ 52

 Make it Right

 19 Gene used symbols to compare numbers.

Why is Gene wrong? Make it right.

9 < 6

 Math at Home: Your child learned how to compare numbers.
Activity: Ask your child to tell three numbers greater than 25 and three numbers less than 80.

Name _____

Order Numbers

Math Words
number line
before
after
between

Learn You can use a number line to help you order numbers.

32 comes before 33.
34 comes after 33.
33 is between 32 and 34.

Try It Write the number that comes just before.

1. 22 , 23, 24
2. ___, 54, 55

Write the number that comes just after.

3. 31, 32, ___
4. 78, 79, ___

Write the number that comes between.

5. 96, ___, 98
6. 49, ___, 51

7. **Write About It!** How do you know which number comes just before 7?

Chapter 6 Lesson 2   ninety-seven **97**

Practice Write the number that comes just before, just after, or between.

1	2	3	4	5	6	7	8	9	10
11	12	13	14	15	16	17	18	19	20
21	22	23	24	25	26	27	28	29	30
31	32	33	34	35	36	37	38	39	40
41	42	43	44	45	46	47	48	49	50
51	52	53	54	55	56	57	58	59	60
61	62	63	64	65	66	67	68	69	70
71	72	73	74	75	76	77	78	79	80
81	82	83	84	85	86	87	88	89	90
91	92	93	94	95	96	97	98	99	100

You can also use a hundred chart to order numbers.

8. 25, 26
9. ___, 14
10. ___, 79
11. ___, 90

12. 61, ___
13. 86, ___
14. 99, ___
15. 13, ___

16. 80, ___, 82
17. 48, ___, 50
18. 59, ___, 61

✶ Algebra • Patterns

Write the missing numbers.

19. Count forward by tens.

    10, 20, 30, _____, 50, 60, _____, 80, 90, 100

20. Count backward by tens.

    100, 90, _____, 70, _____, _____, 40, 30, 20, 10

Math at Home: Your child ordered numbers up to 100.
Activity: Pick a number from 10 to 99. Have your child tell you the number that comes before and after the number.

Name _____

Skip Counting Patterns

ALGEBRA

Learn You can skip count by twos.

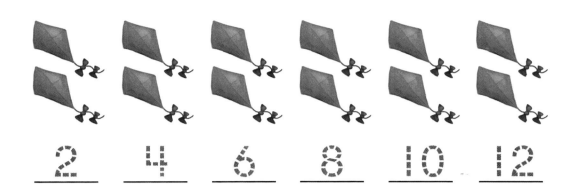

2   4   6   8   10   12

Math Word
skip count

Total
12 kites

Try It Skip count by fives. Write the numbers.

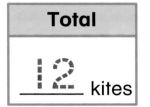

1)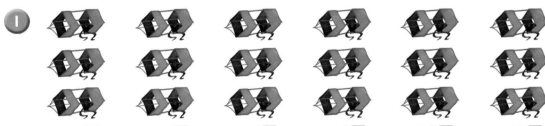

5   ___   ___   ___   ___   ___

Total
___ kites

Skip count by threes. Write the numbers.

2)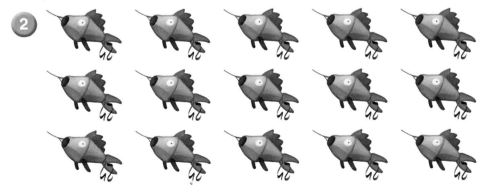

___   ___   ___   ___   ___

Total
___ kites

3) **Write About It!** Look at exercise 1. Tell about the pattern you see.

Chapter 6 Lesson 3

Practice You can use the hundred chart to skip count.

4. Skip count by twos. Color those numbers 🖍.

5. Skip count by threes. Color those numbers 🖍.

6. Skip count by fours. Color those numbers 🖍.

7. Skip count by fives. Color those numbers 🖍.

1	2	3	4	5	6	7	8	9	10
11	12	13	14	15	16	17	18	19	20
21	22	23	24	25	26	27	28	29	30
31	32	33	34	35	36	37	38	39	40
41	42	43	44	45	46	47	48	49	50
51	52	53	54	55	56	57	58	59	60
61	62	63	64	65	66	67	68	69	70
71	72	73	74	75	76	77	78	79	80
81	82	83	84	85	86	87	88	89	90
91	92	93	94	95	96	97	98	99	100

You may color some boxes more than once.

x Algebra • Patterns

8. Write the missing numbers.

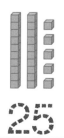 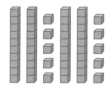 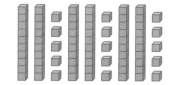 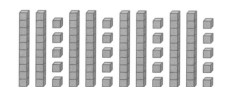

25 _____ _____ _____

Math at Home: Your child counted by twos, threes, fours, and fives to 100.
Activity: Have your child count aloud by twos, threes, fours, and fives to 100.

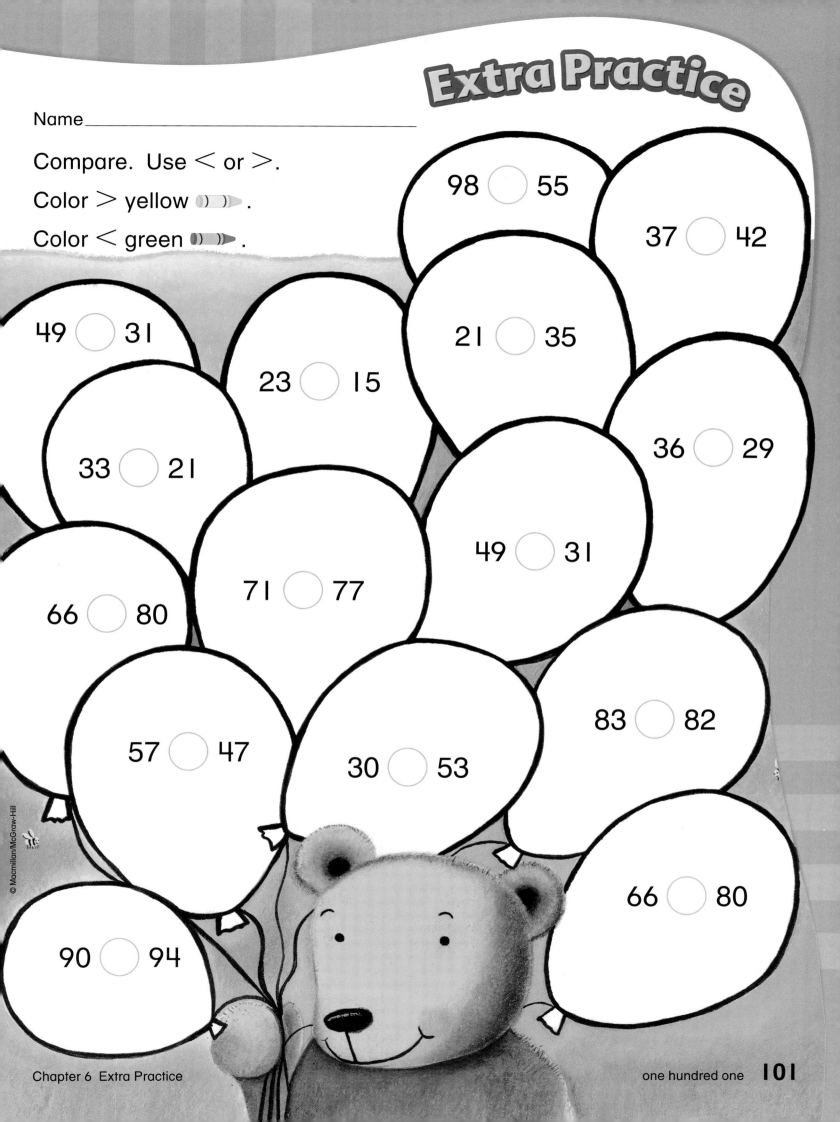

Extra Practice

< is less than   > is greater than   = is equal to

Compare. Use >, <, or =.

1. 4 < 4 + 3
2. 9 ◯ 9 + 2
3. 7 ◯ 7 + 0
4. 6 ◯ 6 + 3
5. 5 ◯ 5 + 0
6. 8 ◯ 8 + 2
7. 7 ◯ 7 − 3
8. 6 ◯ 6 − 2
9. 3 ◯ 3 − 0
10. 5 ◯ 5 − 0
11. 8 ◯ 8 − 2
12. 9 ◯ 9 − 8
13. 8 + 8 ◯ 5 + 5
14. 2 + 2 ◯ 4 + 4
15. 3 + 3 ◯ 6 + 6
16. 7 + 7 ◯ 1 + 1

17. ✏️ Write **About It!**
How did you find the answer for exercise 14? Use pictures or words to explain.

Show Your Work

www.mmhmath.com
LOG ON For more Practice

Math at Home: Your child practiced addition and subtraction facts.
Activity: Have your child tell how he or she found the answer to exercise 13.

one hundred two

Name_____

Ordinal Numbers

Math Word
ordinal numbers

Learn These ordinal numbers tell position.

1st	2nd	3rd	4th	5th	6th	7th	8th	9th	10th
first	second	third	fourth	fifth	sixth	seventh	eighth	ninth	tenth

Try It Circle the correct position of each duck.

1. third / (fourth)
2. first / fifth
3. eighth / ninth
4. third / eighth
5. sixth / seventh
6. fifth / sixth
7. sixth / eighth
8. first / second

9. **Write About It!** How do you know which position is eighth?

Practice Here is a table of ordinal numbers.

1st	first
2nd	second
3rd	third
4th	fourth
5th	fifth
6th	sixth
7th	seventh
8th	eighth
9th	ninth
10th	tenth
11th	eleventh
12th	twelfth
13th	thirteenth
14th	fourteenth
15th	fifteenth
16th	sixteenth
17th	seventeenth
18th	eighteenth
19th	nineteenth
20th	twentieth
21st	twenty-first
22nd	twenty-second
23rd	twenty-third
24th	twenty-fourth
25th	twenty-fifth
26th	twenty-sixth
27th	twenty-seventh
28th	twenty-eighth
29th	twenty-ninth
30th	thirtieth

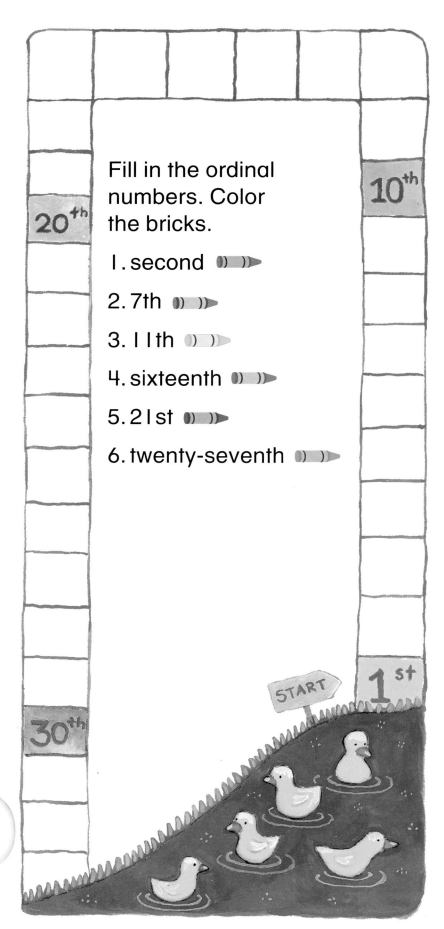

Fill in the ordinal numbers. Color the bricks.

1. second
2. 7th
3. 11th
4. sixteenth
5. 21st
6. twenty-seventh

Follow the pattern to write ordinal numbers to 100th.

Math at Home: Your child used ordinal numbers.
Activity: Have your child put four objects in a line and tell you which is first, second, third, and fourth.

Name_____ **Even and Odd Numbers**

Learn You can make pairs to find if a number is even or odd.

If you make pairs with none left over, the number is even.

If you make pairs with one left over, the number is odd.

Math Words

even

odd

Four is an even number.

Five is an odd number.

Your Turn Use 🧊 to make pairs. Complete the table.

Show	Are any left over?	Circle even or odd.
① 10	no	(even) odd
② 11	_____	even odd
③ 14	_____	even odd
④ 15	_____	even odd
⑤ 17	_____	even odd

⑥ ✏️ **Write About It!** How can you tell if a number is even or odd?

Chapter 6   Lesson 5   one hundred five **105**

Practice Color even numbers 🖍, odd numbers 🖍.

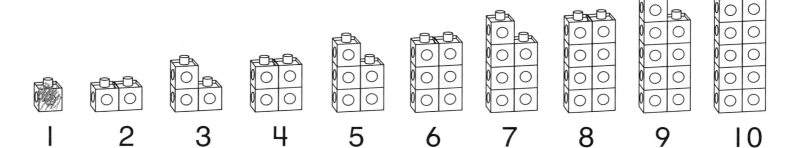

1 2 3 4 5 6 7 8 9 10

Write the next 3 even numbers.

7. 22, ____, ____, ____.
8. 28, ____, ____, ____.
9. 30, ____, ____, ____.
10. 36, ____, ____, ____.
11. 40, ____, ____, ____.
12. 48, ____, ____, ____.

Write the next 3 odd numbers.

13. 27, ____, ____, ____.
14. 39, ____, ____, ____.
15. 41, ____, ____, ____.
16. 45, ____, ____, ____.
17. 53, ____, ____, ____.
18. 59, ____, ____, ____.

Problem Solving Reasoning

19. **15 25 35**

How does the 5 in the ones place help you know that these are odd numbers? _____

Math at Home: Your child learned about even and odd numbers.
Activity: Say a number from 1 to 20. Have your child tell if it is odd or even.

106 one hundred six

Name _____

Problem Solving Strategy

Use Logical Reasoning

You can use logical reasoning to help you solve problems.

Jan sits behind Ann.
Ann sits behind Kim.
Who sits in front?

Read

What do I already know?   _____ sits behind Ann.

_____ sits behind Kim.

What do I need to find out?   _____

Plan

I can use the clues to solve the problem.
I can draw pictures to show the clues.

Solve

I can carry out my plan.

First Clue:    Second Clue:

_____   _____

Who sits in front? _____

Look Back

Does my answer make sense?   Yes   No

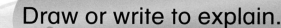

Use logical reasoning to solve.

Draw or write to explain.

1. Three children fly kites.
Lin is next to Jim.
Kay is next to Lin.
Who is in the middle?

___Lin___

2. Three boys ride bicycles.
Pat bicycles behind Bill.
Bill bicycles behind Rob.
Who rides in front?

_____

3. Pam, Deb, and Sue play ball.
Pam's ball is the smallest.
Deb's ball is not the largest.
Whose ball is the largest?

_____

4. There are four boys in
line to buy popcorn.
Ray is third. Mike is after Ray.
Jake is before Dan.
Dan is second. Who is first?

_____

Math at Home: Your child solved problems by using logical reasoning.
Activity: Ask your child to describe how he or she solved problem 4.

Name_____

Game Zone

Practice at School ★ Practice at Home

What's My Number?

▶ Play with a partner. Take turns.

▶ Write a number. Give a clue, such as, "My number is less than 30."

▶ Your partner can ask 4 questions to guess your number.

▶ Answer questions with yes or no.

▶ The player who guesses more numbers correctly wins.

 2 players

1	2	3	4	5	6	7	8	9	10
11	12	13	14	15	16	17	18	19	20
21	22	23	24	25	26	27	28	29	30
31	32	33	34	35	36	37	38	39	40
41	42	43	44	45	46	47	48	49	50

Technology Link

Place Value • Calculator

You can use a to build a number.

Show 1 ten 4 ones.

Press.

$\underline{10} + \underline{4} = \underline{14}$

Show 2 tens 3 ones.

Press.

$\underline{20} + \underline{3} = \underline{23}$

---

1. Show 1 ten 5 ones.
   Press.

2. Show 2 tens 4 ones.
   Press.

3. Show 3 tens 8 ones.
   Press.

You can use a pencil or a calculator.

Chapter 6 Review/Test

Name_____

Compare. Use <, >, or =.

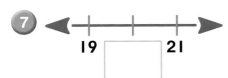
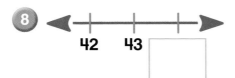
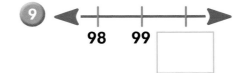

① 37 ◯ 52   ② 49 ◯ 41   ③ 70 ◯ 44

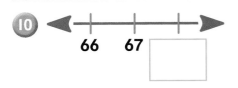
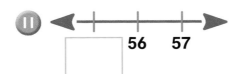
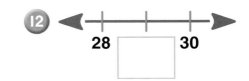

④ 16 ◯ 61   ⑤ 28 ◯ 28   ⑥ 56 ◯ 60

Write each missing number.

⑦ 19 ___ 21   ⑧ 42 43 ___   ⑨ 98 99 ___

⑩ 66 67 ___   ⑪ ___ 56 57   ⑫ 28 ___ 30

Skip count by fives.

⑬ 5 10 ___ ___ ___ ___ ___ ___

Tell if the number is even or odd.

⑭ 19 _____   ⑮ 30 _____   ⑯ 65 _____

⑰ 51 _____   ⑱ 96 _____   ⑲ 72 _____

⑳ Tom is older than Rob. Rob is not older than Sam. Sam is younger than Tom. Who is the oldest? Use words or pictures to explain.

_____

Spiral Review and Test Prep
Chapters 1–6

Choose the best answer.

1 How many are there in all?

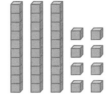

28	32	38	83
○	○	○	○

2 Which number sentence matches this picture?

$3 + 2 = 5$	$2 + 0 = 2$	$3 + 0 = 3$	$3 + 3 = 6$
○	○	○	○

3 Write the number word for 30.

_____

4 Write the missing number. $8 + 4 = 4 +$ _____

Solve.

5 There are 7 red cars, 5 blue cars, and 3 black cars. How many cars are there in all? Draw a picture and explain how you solved it. Show the number sentence used.

_____

_____

www.mmhmath.com
For more Review and Test Prep

UNIT 2 CHAPTER 7

Money

A Visit to the Fair

I'm saving up my money,

And I'm going to the fair.

All my nickels, dimes, and qu[arters]

I can't wait to spend them the[re]

A quarter buys some popcorn[,]

A dime buys a balloon,

And fifty cents will get me

On a space ride to the moon!

Math at Home

Dear Family,

I will learn to show and count money amounts in Chapter 7. Here are my math words and an activity that we can do together.

Love, _____

Home Activity

Place some pennies, nickels, dimes, and quarters in a bag. Take turns with your child.

Reach into the bag and take out 1 coin. No peeking. Tell one thing about the coin, for example, its name, value, size, or color.

Books to Read

Look for these books at your local library and use them to help your child learn money concepts.

- **Alexander, Who Used to Be Rich Last Sunday** by Judith Viorst, Atheneum Books, 1978.
- **Yard Sale** by James Stevenson, Greenwillow Books, 1996.
- **The Coin Counting Book** by Rozanne Lanczak Williams, Charlesbridge, 2001.

www.mmhmath.com
For Real World Math Activities

one hundred fourteen

Name_____

Count Coin Collections

Learn Is there enough money to buy the toy train?

Find the total amount. Start with the coins that have the greatest value. You can skip-count.

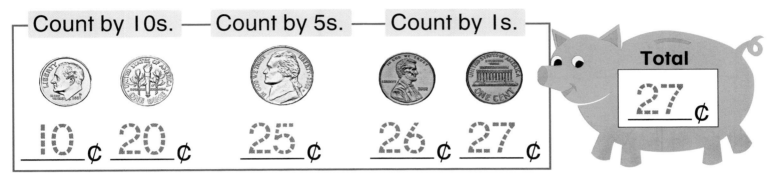

Count by 10s.	Count by 5s.	Count by 1s.	Total
10¢ 20¢	25¢	26¢ 27¢	27¢

There is not enough money to buy the train.

Your Turn Put 5 coins in this bank each time. Find the total amount. Is there enough money to buy the train?

The toy train costs 30¢.

1.

2. **Write About It!** Draw 5 coins. Do you have enough money to buy the train?

Practice Count to find the total amount.
Do you have enough money to buy the toy?
Color the toys you can buy.

Toy	Coins	Total Amount
3. horse 30¢	dime, nickel, dime, nickel	30¢
4. butterfly 80¢	nickel, dime, dime, nickel, dime, dime, dime	____¢
5. sailboat 35¢	nickel, nickel, nickel, nickel, nickel	____¢
6. dolphin 30¢	dime, nickel, dime, dime	____¢
7. bicycle 80¢	dime, dime, dime, penny, dime, dime, dime	____¢

Problem Solving Critical Thinking

Show Your Work

8. Josie has some nickels and dimes. She buys a ring for 20¢. How many different ways can Josie pay for her ring? Show all the ways.

Math at Home: Your child counted groups of coins to 99¢.
Activity: Give your child a group of coins to count. Help your child arrange the coins in order starting with the greatest value.

118 one hundred eighteen

Name_____   **Money and Place Value**

Learn You can use money to show place value.

2	4
tens	ones

24 = 2 tens 4 ones

A penny is the same as 1 one. A dime is the same as 1 ten.

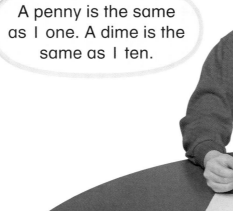

Your Turn Use coins. Write how many tens and ones.

1.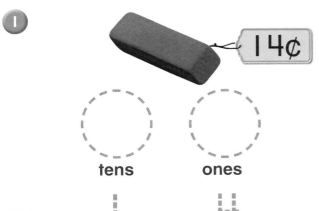

   tens   ones

   14 = __1__ ten __4__ ones

2.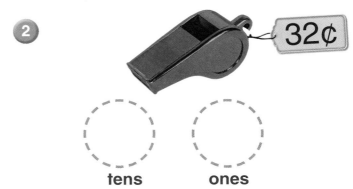

   tens   ones

   32 = ____ tens ____ ones

3.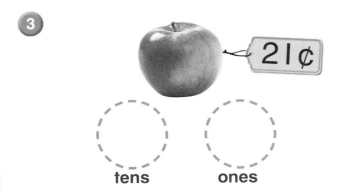

   tens   ones

   21 = ____ tens ____ one

4.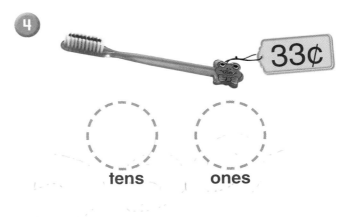

   tens   ones

   33 = ____ tens ____ ones

5. **Write About It!** How many tens and ones are in 57¢?

Practice Use money to show place value.

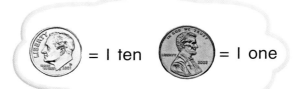
= 1 ten   = 1 one

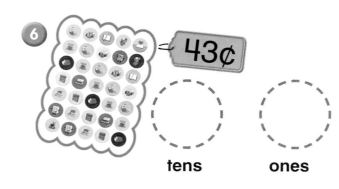
43¢
tens   ones

43 = ___ tens ___ ones

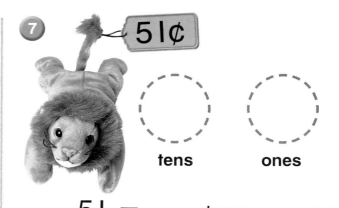
51¢
tens   ones

51 = ___ tens ___ one

⑧ 76 = ___ tens ___ ones   ⑨ 39 = ___ tens ___ ones

⑩ 94 = ___ tens ___ ones   ⑪ 28 = ___ tens ___ ones

⑫ 42 = ___ tens ___ ones   ⑬ 81 = ___ tens ___ one

⑭ 19 = ___ ten ___ ones   ⑮ 66 = ___ tens ___ ones

⑯ 52 = ___ tens ___ ones   ⑰ 23 = ___ tens ___ ones

x Algebra • Patterns

What comes next in each pattern?

⑱ 12¢, 13¢, 14¢, ___¢, ___¢, ___¢, ___¢.

⑲ 30¢, 40¢, 50¢, ___¢, ___¢, ___¢, ___¢.

Math at Home: Your child related dimes and pennies to tens and ones.
Activity: Cut from a supermarket flyer pictures of things that sell for under $1.00. Give your child pennies and dimes and ask him or her to show how many tens and ones are in the price of each item.

one hundred twenty

Name_____

Problem Solving Practice

Solve.

1. Dan has 3 . Does he have enough to buy  for 70¢?

   _____

2. Jane has 3 🪙. Pam has 2 🪙. How much money do they have in all?

   _____¢

Write a Story!

3. You have 2 quarters, 2 dimes, and 3 nickels. How much money do you have in all? Write a story about how you would spend it at a fair.

   _____¢

   _____
   _____
   _____

4. Jon spends 2 quarters on a game. Then he buys a pretzel for 4 dimes. How much money did he spend in all?

   _____¢

5. A cup of lemonade costs 50¢. Tanya pays for it with 2 coins. What 2 coins does she use?

   _____

Chapter 7 Problem Solving Practice

Writing for Math

Write a money story about the picture.
Use the words on the cards to help.

money   enough

40¢

Think

Count how much money the boy has.

____¢   ____¢   ____¢   ____¢   ____¢   ____¢

Solve

I can write my story now.

_____

_____

_____

_____

Explain

I can tell you how my story matches the picture.

_____

_____

_____

UNIT 2 CHAPTER 8

Using Money

Money Rhymes

Twenty-five cents,
Money that rhymes.
Take one nickel,
Add two dimes.

Three fat nickels,
One thin dime
Makes twenty-five cents
Every time.

Five fat nickels,
No thin dimes
Makes twenty-five cents
Any time.

Math at Home

Dear Family,

I will learn about using dollars and cents in Chapter 8. Here are my math words and an activity that we can do together.

Love, _____

My Math Words

dollar:

cent sign:

dollar sign:

$1.00

decimal point:

$1.00

Home Activity

Put some price tags below 50¢ on some household objects.

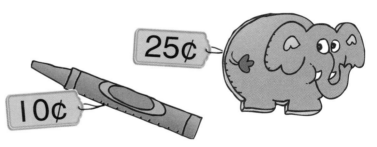

Place objects with price tags on a table. Ask your child to use coins to show the price.

Books to Read

In addition to these library books, look for the Time For Kids math story that your child will bring home at the end of this unit.

- **A Chair For My Mother** by Vera Williams, Morrow, William, and Company, 1984.
- **Dollars and Cents For Harriet** by Betsy and Giulio Maestro, Crown Publishers, 1988.
- **Time For Kids**

www.mmhmath.com
For Real World Math Activities

Name _____   **Make Change**

Learn Darcy buys a plant for 48¢.
She gives Jason 50¢.
How much change does Darcy get back?

Start at 48¢. Count up to 50¢. Say 49¢, 50¢.

__49¢__ , __50¢__

48¢   The change is __2¢__.

Your Turn Count up from the price to find the change.
Use coins. Draw to show the change.

You Have	Price	Count up to find your change.
1. quarter	22¢	1¢ 1¢ 1¢ __23¢__ __24¢__ __25¢__ The change is __3¢__.
2. quarter, dime, nickel	38¢	_____ _____ The change is _____.

3. **Write About It!** How can you count up to make change? Give an example.

Chapter 8 Lesson 4   one hundred thirty-nine **139**

Practice Use coins to show the change.

Count on from the price to find the change.

You Have	Price	Count up to find your change.
④ (quarter)	basket of flowers 24¢	25¢ The change is 1¢.
⑤ (two dimes)	potted flowers 17¢	_____ _____ _____ The change is _____.
⑥ ($1 bill)	ivy plant 98¢	_____ _____ The change is _____.

Problem Solving — Estimation

Show Your Work

⑦ Mike has 2 quarters and 20 pennies. Do you think he has more than $1.00 or less than $1.00? Draw pictures or use numbers to check.

Math at Home: Your child learned to count up from a price to make change.
Activity: Give your child a price less than a dollar and an amount given as payment and have your child count up to make change.

Name _____

Problem Solving Strategy

Act It Out

Sometimes you can act out a problem.

Jack has 2 quarters. He buys a pinwheel for 29¢. How much change should he get?

Read

What do I know already?

Jack has _____.

One pinwheel costs _____.

What do I need to find out?

_____

Plan

I need to count on from the price to find the change. I can use coins to act it out.

Solve

Start with a penny. Count from the price. Write the change.

_____ _____ _____   _____ change

Look Back

Does my answer make sense?   Yes   No

Chapter 8   Lesson 5   one hundred forty-one **141**

Read Plan Solve Look Back

Use coins. Act it out to solve.

Draw or write to explain.

① Judy has a dollar bill. She buys a book for 92¢. How much change should she get?

~~8¢~~ change

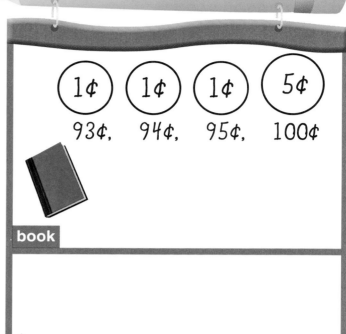

(1¢) (1¢) (1¢) (5¢)
93¢, 94¢, 95¢, 100¢

book

② Bob has 1 half dollar and 3 dimes. He buys a lemonade for 71¢. How much change should he get?

_____ change

lemonade

③ Kay has 4 quarters. She buys a pretzel for 81¢. How much change should she get?

_____ change

pretzel

④ Jose has 3 quarters. He buys a yo-yo for 53¢. How much change should he get?

_____ change

yo-yo

Problem Solving

Math at Home: Your child solved problems by using the strategy Act It Out.
Activity: Give your child a price less than one dollar. Have your child show you how to find the change from a dollar.

142 one hundred forty-two

Name _____

About 10

About 15

About 25

About 70

Write the numbers in order from least to greatest.

____
____
____
____

Fold down

How Many Marchers?

Children march in the big parade.

Listen to the marching band. More than 70 marchers look so grand.

About 15 dancers dance in the street.

About 25 cheerleaders move to the beat.

Name _____

Linking Math and Science

Numbers and Flowers

Some animals bring pollen from flower to flower. Pollen makes seeds grow inside flowers. Petals are parts of flowers. They attract animals to flowers.

Science Words
pollen
petal

Circle the word to complete each sentence.

1. The part of a flower that makes seeds grow is _____.

   pollen   petal

2. A flower part that attracts animals is the _____.

   petal   pollen

Unit 2 Linking Math and Science — one hundred forty-seven 147

What to Do

- **Observe** Use . Show how many petals are on each flower.

- **Classify** Write the number of petals. Tell if it is even or odd.

This flower has an odd number of petals.

You Will Need

cubes

③

Flower				
Number of Petals	5			
Odd or Even?	odd			

Problem Solving

Solve.

④ **Put in Order** Write the numbers of petals in order from least to greatest.

_____ , _____ , _____ , _____

⑤ **Compare** Compare the number of petals. Use < or >.

 ⑥ **Investigate** What are some other parts of a plant?

_____

 Math at Home: Your child used numbers to investigate differences among flowers.
Activity: Repeat this activity with your child. Find pictures of other flowers in field guides or magazines.

Name _____

Unit 2
Study Guide and Review

Math Words
Draw lines to match.

1. 5, 11, 19
2. 4 quarters
3. seventh, eighth, ninth

one dollar

ordinal numbers

odd numbers

Skills and Applications
Numbers to 100 (pages 77-86, 95-108)

Examples

Regroup ones as tens and ones.

26 ones → 2 tens 6 ones

4. 63 = ____ tens ____ ones

5. 39 = ____ tens ____ ones

6. 81 = ____ tens ____ one

Use <, >, or = to compare numbers.

34 is less than 37.   34 < 37

34 is greater than 31. 34 > 31

34 is equal to 34.   34 = 34

7. 97 ◯ 78
8. 55 ◯ 63
9. 23 ◯ 23
10. 64 ◯ 69

Unit 2 Study Guide and Review

one hundred forty-nine **149**

Skills and Applications

Money (pages 115-124, 133-140)

> Start with the coins that have the greatest value.

Examples

Count to find the total amount.

50¢,   60¢,   65¢

Total
65¢

11.

_____ ¢ _____ ¢ _____ ¢

Draw coins to show the change.

You pay

1¢ 1¢ 1¢ 1¢ 1¢

Your change is 5¢.

12. You pay .

 8¢

Your change is _____ ¢.

Problem Solving — Strategy

(pages 107-108, 141-142)

Show Your Work

13. Jane lives on the 10th floor.
Max lives 3 floors below Jane.
Kim lives 5 floors above Max.
Use a picture to show where
Jane, Max, and Kim live.

Draw a picture to solve.

Math at Home: Your child practiced using numbers to 100 and counting dollar bills and coins.
Activity: Have your child use these pages to review numbers to 100 and counting money.

Name_____

Unit 2 Performance Assessment

Number Riddle

I have 6 tens.

I am less than 69.

I am an odd number.

What number could I be?

You Will Need

- Use ▭ and ▫ to show your answer.

- Use words or pictures to tell why your answer makes sense.

- Then make up your own number riddle.

Show your work.

You may want to put this page in your portfolio.

Unit 2
Enrichment

Different Ways to Show Numbers

Here are different ways to show 24.

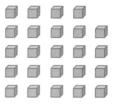 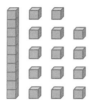 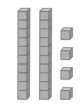 

24 ones = 1 ten 14 ones = 2 tens 4 ones

Draw lines to match.

42	13 ones
63	5 tens 28 ones
24	3 tens 12 ones
96	4 tens 11 ones
13	9 tens 6 ones
35	3 tens 5 ones
51	5 tens 13 ones
78	2 tens 4 ones

UNIT 3

CHAPTER 9

Telling Time

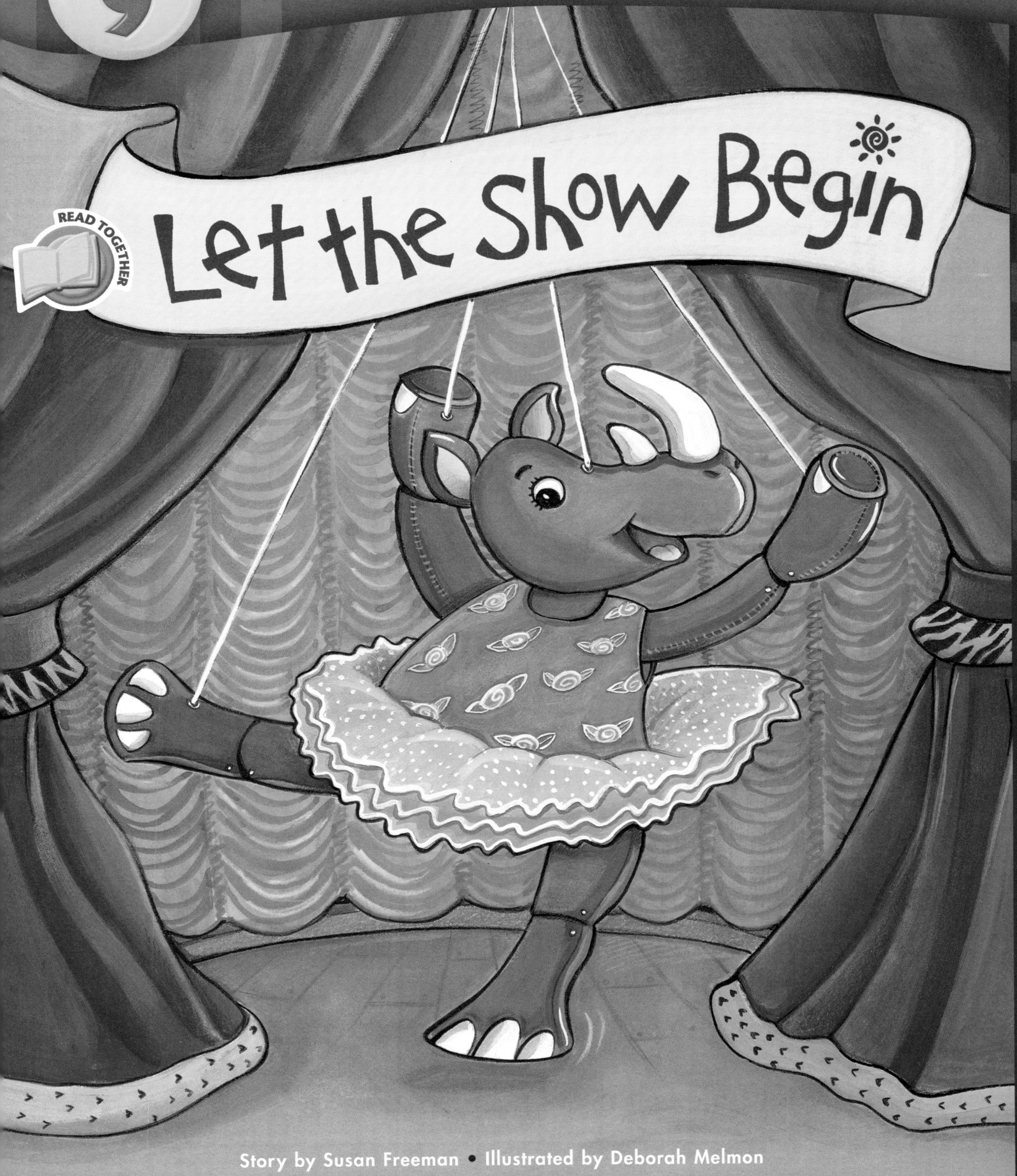
Let the Show Begin
READ TOGETHER

Story by Susan Freeman • Illustrated by Deborah Melmon

The juggling elephants start the show.
They jump, they spin, and eggs they throw!
Look at the clock.

What time is it now? _____ o'clock

The zebras zag, the zebras zig.
They do a funny zebra jig.
Look at the clock.

What time is it now? _____ : _____

The rhinos race, the rhinos run.
They have lots of rhino fun.
Look at the clock.

Is it before or after 4 o'clock? _____

All the animals sing and dance, as across the stage they prance! Look at the clock.

What time is it now? ____ : ____

153D

Math at Home

Dear Family,
  I will learn to tell time in Chapter 9. Here are my math words and an activity that we can do together.
  Love, _____

My Math Words

hour :
60 minutes

 8:00

minute :
60 seconds

  8:01

half hour :
30 minutes

 8:30

LOG ON
www.mmhmath.com
For Real World Math Activities

Home Activity

Ask your child to name an activity that he or she can do in one minute. Time your child while he or she does the activity to see how long it actually takes.

  7:00

Books to Read

Look for these books at your local library and use them to help your child learn to tell time.

- **Clean Your Room, Harvey Moon!** by Pat Cummings, Aladdin, 1991.
- **The Grouchy Ladybug** by Eric Carle, HarperCollins, 1986.
- **Game Time!** by Stuart J. Murphy, HarperCollins, 2000.

Name_____

Time to the Quarter Hour

Learn Every 15 minutes the ducks sing a song.
The ducks sing 4 songs in one hour.
Write each time.

Math Word
quarter hour

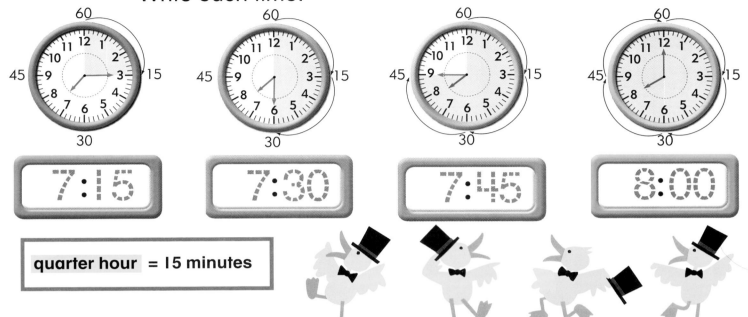

quarter hour = 15 minutes

Try It Write each time.

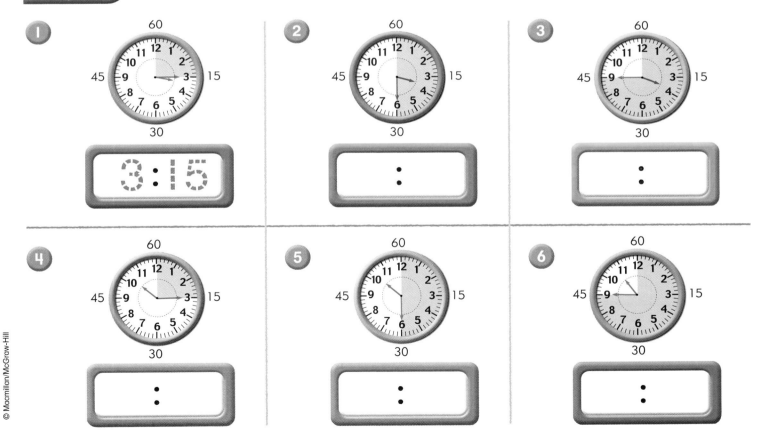

7. **Write About It!** How many quarter hours are in 1 hour? How do you know?

Chapter 9 Lesson 3

Practice Write each time.

8.  4:45

9.  :

There are 4 quarter hours in 1 hour.

10.  :

11.  :

12.  :

Draw the minute hand to show each time.

13.  7:15

14.  7:30

15.  7:45

x Algebra • Patterns

Extend the clock patterns.

16. 8:10   8:20   8:30   :   :

17. 2:00   2:05   2:10   :   :

Math at Home: Your child practiced telling time to the quarter hour.
Activity: Set a clock to show 5:00. Ask your child what time it will be 15 minutes later.

Name_____   **Time Before and After the Hour**

Learn You can tell time in different ways.

50 minutes after 8
or 10 minutes before 9

quarter after 10
or 15 minutes after 10

Try It Write each time more than one way.

1.

   45 minutes after 1

   15 minutes before 2

2.

   ____ minutes after ____

   quarter after ____

3.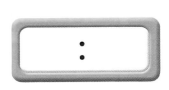

   ____ minutes after ____

   ____ minutes before ____

4.

   ____ minutes after ____

   quarter after ____

5. **Write About It!** Write the time you eat lunch.
   Write it using words and numbers.

Chapter 9 Lesson 4   one hundred sixty-one **161**

Practice Write each time more than one way.

You can tell time in different ways.

6)

<u>30</u> minutes after <u>2</u>

7)

____ minutes after ____

8)

____ minutes after ____

quarter after ____

9)

____ minutes after ____

____ minutes before ____

Problem Solving — Visual Thinking

Draw hour and minute hands to show each time.

10)

11)

12)

13)

Math at Home: Your child practiced telling time in different ways.
Activity: Ask your child to tell the current time in different ways. Repeat throughout the day.

Name_____

Problem Solving Practice

Solve.

1. Bob leaves for school at 8:30. Draw the hands on the clock to show that time.

2. Bob plays soccer after school. He gets home at 4:00. Draw the hands on the clock to show that time.

3. Sarah starts an ice skating lesson at 5:15. She skates for one hour and then goes home. Draw the hands on the clock to show when Sarah goes home.

4. Joe's pet shop opens at 9:00. It is open for eight and one-half hours. What time does the pet shop close? Draw the hands on the clock to show that time.

 Write a Story!

5. Your family leaves on a camping trip at 6:30. It takes 2 hours to get to the lake and 45 minutes to set up your tent! What time is it then? Write a story about your trip.

_____

_____

Chapter 9 Problem Solving Practice

Writing for Math

 Draw hour and minute hands to show the time you go to school. Write two different ways to read the time. Explain your work.

_____

_____

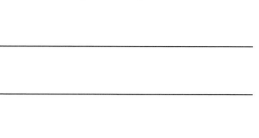

Think

At what time do I go to school?

_____

Solve

I can write two ways to read the time.

_____

_____

Explain

I can tell you how I know.

_____

_____

_____

Chapter 9
Review/Test

Name _____

Write each time.

Draw hands to show each time.

10:15

2:55

7:25

Write each time more than one way.

_____ minutes after 1

quarter after _____

45 minutes after _____

15 minutes before _____

30 minutes after _____

30 minutes before _____

 Tanya leaves school at quarter after two. Write the time on the clock.

Chapter 9 Review/Test

one hundred sixty-seven **167**

Spiral Review and Test Prep
Chapters 1–9

Choose the best answer.

1 Which number has a 5 in the tens place?

    5   23   25   51
  ◯   ◯   ◯   ◯

2 Which number makes the number sentence true?

$14 = \Box + 9$

    4   5   9   10
  ◯   ◯   ◯   ◯

3 How much money is there?

____ ¢

4 Complete the pattern.

2:00, 2:15, 2:30, ____, ____, ____, ____

5 Donna saw 8 butterflies. The next day Rob saw 2 fewer than Donna. How many butterflies did they see in all? Solve with pictures or numbers.

UNIT 3
CHAPTER 10

Time and Calendar

Going to Bed

This little boy is going to bed.
Down on the pillow he lays his head,
Wraps himself in the cover tight.
This is the way he sleeps all night.

Morning comes, he opens his eyes.
Back with a toss the cover flies.
Up he jumps, is dressed and away,
Ready for work, ready for play.

Math at Home

Dear Family,

I will learn more about time and how to read a calendar in Chapter 10. Here are my math words and an activity that we can do together.

Love, _____

My Math Words

A.M.: time between midnight and noon

P.M.: time between noon and midnight

calendar:

Home Activity

Ask your child what time he or she got up this morning. Write the time. Have your child set the hands of a clock to show the time.

Repeat using times of other daily events, such as dinnertime and bedtime.

Books to Read

Look for these books at your local library and use them to help your child learn about the calendar and other time concepts.

- **Play Date** by Rosa Santos, Kane Press, 2001.
- **The Turning of the Year** by Bill Martin Jr., Harcourt Brace and Company, 1970.
- **One Lighthouse, One Moon** by Anita Lobel, Greenwillow Books, 2000.

www.mmhmath.com
For Real World Math Activities

Name_____

A.M. and P.M.

Learn The hours from midnight until noon are labeled A.M.
The hours from noon to midnight are labeled P.M.

Math Words
A.M.
P.M.

12:00 P.M. is noontime.
12:00 A.M. is midnight.

 8:00 A.M.

 8:00 P.M.

Try It Circle the best time for each picture.

1.  6:30 A.M. or (6:30 P.M.)

2. 7:15 A.M. or 7:15 P.M.

3.  8:00 A.M. or 8:00 P.M.

4.  8:30 A.M. or 8:30 P.M.

5. **Write About It!** Think about something you do at 11:30 A.M. Write about it.

Practice Write A.M. or P.M. for each picture.

6. Go to school.

   A.M.

7. Get ready for bed.

   _____

8. Do homework.

   _____

9. Eat dinner.

   _____

Circle the best time for each picture.

10.

    2:00 A.M.   2:00 P.M.

11.

    11:00 A.M.   11:00 P.M.

Spiral Review and Test Prep

12. Which set of numbers shows least to greatest?

    12, 8, 21   12, 21, 19   22, 12, 10   12, 21, 22
    ○   ○   ○   ○

13. Ted had $1.25. He lost one coin. Now he has $1.00. Which coin did he lose?

    penny   nickel   dime   quarter
    ○   ○   ○   ○

Math at Home: Your child learned to tell time using A.M. and P.M.
Activity: Make a list of daily events. Have your child tell whether the events happen in the A.M. or P.M.

Name_____ **Elapsed Time**

Learn
Use a 🕐 to help you find out how much time has passed.

The baseball game lasts 2 hours.

The game starts at 11:00 A.M.

The game ends at 1:00 P.M.

Your Turn
Complete the table. Use a 🕐 to help you.

Activity	Start Time	End Time	How long does it take?
1. Have a bike race.	10:00 A.M.	12:00 P.M.	2 hours
2. Play a soccer game.	___:___ P.M.	___:___ P.M.	___ hour

3. ✏️ **Write About It!** If a party starts at 3:00 P.M. and ends at 6:00 P.M., how many hours have passed?

Chapter 10 Lesson 2   one hundred seventy-three **173**

Practice Complete the table. Use a  to help you.

> A.M. means from midnight until noon.
> P.M. means from noon until midnight.

Activity	Start Time	End Time	How long does it take?
④ We went roller skating.	9:00 A.M.	12:00 P.M.	3 hours
⑤ We played basketball.	___:___ A.M.	___:___ P.M.	___ hours
⑥ We had a track meet.	___:___ A.M.	___:___ P.M.	___ hours

Problem Solving — Number Sense

⑦ Ken used the computer from 9:30 until 11:00. Draw the hands. How much time has passed?

Start 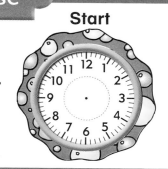   End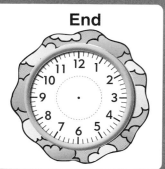

Math at Home: Your child learned how to tell how much time has passed.
Activity: Tell your child at what time you start to prepare dinner and at what time you finish. Ask how long it took you. Repeat with other activities.

Extra Practice

Name _____

Add. Extend the patterns.

1)
 0   1   2   3   ☐
+8   +7   +6   +5   +☐
 ☐   ☐   ☐   ☐   ☐

2)
 7   6   5   4   ☐
+2   +3   +4   +5   +☐
 ☐   ☐   ☐   ☐   ☐

3)
 9   8   7   6   ☐
+1   +2   +3   +4   +☐
 ☐   ☐   ☐   ☐   ☐

4)
 4   5   6   7   ☐
+7   +6   +5   +4   +☐
 ☐   ☐   ☐   ☐   ☐

5) **Write About It!** What pattern do you see in Exercise 4?

Chapter 10 Extra Practice   one hundred seventy-five **175**

Extra Practice

Count each group of coins.
Write the total amount.
Then compare. Use >, <, or =.

is greater than >
is less than <
is equal to =

_____ ¢   _____ ¢

_____ ¢   _____ ¢

_____ ¢   _____ ¢

_____ ¢   _____ ¢

 www.mmhmath.com
For more Practice

 Math at Home: Your child practiced addition facts and counting money.
Activity: Have your child show you how to count a group of mixed coins.

176 one hundred seventy-six

Name_____ **Time Relationships** ALGEBRA

Learn Use the best units to estimate and measure time.

Time Relationships	
1 minute = 60 seconds	1 week = 7 days
1 hour = 60 minutes	1 year = 12 months
1 day = 24 hours	

Try It Circle the best unit to measure time for each event.

1. to brush your teeth

   (minutes)   hours

2. to jump rope

   minutes   days

3. to build a house

   hours   months

4. to bake a cake

   hour   week

5. **Write About It!** How many days are in 2 weeks? How do you know?

Chapter 10 Lesson 4

Practice Circle the best units to measure the time for each event.

1 hour = 60 minutes
1 day = 24 hours
1 week = 7 days
1 month = 4 weeks
1 year = 12 months or 52 weeks

6. to do your homework

(minutes)   weeks

7. to build a playground

hours   weeks

8. to go to the store

hour   week

9. to eat lunch

minutes   days

Problem Solving Number Sense

10. Amy visited her aunt for 1 month. How long did she visit? Circle the answer.

4 days   4 weeks

Name _____

Problem Solving Strategy

Use a Model

You can use a calendar to help you solve problems.

Mim is coming home on November 14. Sabrina has planned a party for Mim on the next Sunday. What is the date of the party?

November

Sunday	Monday	Tuesday	Wednesday	Thursday	Friday	Saturday
		1	2	3	4	5
6	7	8	9	10	11	12
13	14	15	16	17	18	19
20	21	22	23	24	25	26
27	28	29	30			

Read

What do I already know?   Mim is coming on November _____.

The party is on the next _____.

What do I need to find? _____

Plan

I need to find the date of the first Sunday after November 14. I can read the calendar.

Solve

I can carry out my plan. The date of the Sunday after November 14 is _____.

Look Back

Does my answer make sense?   Yes   No

How do I know? _____

Chapter 10 Lesson 5

one hundred eighty-one 181

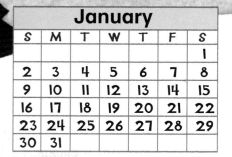

Write the correct date. Use the calendar to solve the problems.

1. Marc goes to the dentist on the third Tuesday in January. What is the date?

   Tuesday, January 18

2. Lucy starts her dance class on the first Monday in February. What is the date?

   Monday, February ____

3. The students will see the school play on the second Thursday in March. What is the date?

   Thursday, March ____

4. Ken is going skating with his family the third Saturday in February. What is the date?

   Saturday, February ____

5. The last day of January falls on what day of the week?

   ____

6. The first day of February falls on what day of the week?

   ____

Math at Home: Your child used a calendar to solve problems.
Activity: Use a calendar at home. Ask your child to find the third Tuesday or second Wednesday of the current month. Then ask him or her to find the date of the following Monday.

182 one hundred eighty-two

Name_____

Game Zone

Practice at School ★ Practice at Home

Busy Bees

▶ Cover each clock with a counter.
▶ You and your partner take turns.
▶ Remove 2 counters.
▶ Keep counters if the times match.
▶ Put counters back if they do not.
▶ The player with more counters wins.

 2 players

You Will Need

14 ●

Chapter 10 Game Zone

one hundred eighty-three 183

Technology Link

Time • Calculator

You can use a  to find the time.

Number of Weeks	Number of Days
1	7
2	?

Press (On/Off) [7] [+] [7] [=] → 14

The answer is __14__ days.

Use the calculator. Complete the pattern.

Number of Weeks	Number of Days
1	7
2	14
3	21
4	
5	
6	
7	
8	

Think:
7 + 7 + 7 = 21

184 one hundred eighty-four · Chapter 10 Technology Link

Chapter 10 Review/Test

Name _____

Complete the table.

Start Time	End Time	How much time passed?
1. (clock showing 2:00) ___:___ P.M.	(clock showing 5:00) ___:___ P.M.	___ hours
2. (clock showing 11:00) ___:___ A.M.	(clock showing 2:00) ___:___ P.M.	___ hours

Answer the questions.

3. What day comes after Monday?

   _____

4. What is the first month of the year?

   _____

Use a model to solve.

5. Joe read a book from 11:30 A.M. until 1:30 P.M. How much time passed?

   _____

Start

End

Spiral Review and Test Prep
Chapters 1–10

Choose the best answer.

1 What is the missing number?

Input	1	2	3	4	5
Output	3	6	9	?	15

10 ○   11 ○   12 ○   21 ○

2 What is the value of the digit 4?

40¢ each

4 ones ○   4 tens ○   48 ones ○   40 tens ○

3 Jeff buys a yo-yo. He pays with . How much change should he get?    48¢

2¢ ○   12¢ ○   48¢ ○   50¢ ○

4 What fact is missing from this fact family?

8 + 5 = 13   13 − 5 = 8   13 − 8 = 5   _____

5 Write a related addition fact for 8 + 9 = 17.

___ + ___ = 17

www.mmhmath.com
For more Review and Test Prep

186 one hundred eighty-six

UNIT 3 CHAPTER 11

Data and Graphs

Under the Weather

We had two days of rain,
Then three days of snow—
I didn't wear boots
When I dressed up to go

Splashing in puddles
And riding my sled,
As raindrops and snowflakes
Came down on my head.

Now one day of sunshine—
The sky's clear and blue!
But I'm stuck in bed
With a cold. AH-AH-CHOO!

Math at Home

Dear Family,

I will learn how to read and make graphs in Chapter 11. Here are new vocabulary words and an activity that we can do together.

Love, _____

My Math Words

picture graph :

tally mark :

A mark used to record data

| = 1 ||||| = 5

bar graph :

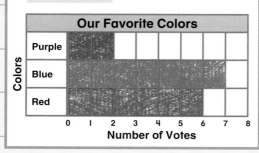

www.mmhmath.com
For Real World Math Activities

Home Activity

Make number cards for 1 to 20.

Shuffle and divide the cards in 2 stacks. Each of you turns over a card. The person who drew the greater number wins both cards. Repeat until all cards are drawn. The winner has the most cards.

Books to Read

Look for these books at your local library and use them to help your child learn about data and graphs.

- **Dave's Down-to-Earth Rock Shop** by Stuart J. Murphy, HarperCollins, 2000.
- **Bart's Amazing Charts** by Dianne Ochiltree, Scholastic, 1999.
- **Hannah's Collections** by Marthe Jocelyn, Dutton Children's Books, 2000.

Chapter 11 Review/Test

Name _____

Use data from the pictograph to answer questions 1–2.

1 How many children voted for jays?

_____

2 How many more children voted for robins than crows?

_____

Favorite Birds

Crow	☺
Robin	☺ ☺ ☺ ☺
Jay	☺ ☺ ☺ ☺ ☺ ☺

Key: Each ☺ stands for 2 votes.

Use data from the chart to make a bar graph. Then answer questions 3–5.

Pat's Book Collection

Kind of Book	Tally
Poems	||
Animals	||||
Sports	|||||

Pat's Book Collection

Kind of Book: Poems, Animals, Sports
Number of Books: 0 1 2 3 4 5

3 How many sports books does Pat have?

_____

4 Which kind of book does Pat have the greatest number of?

_____

5 Which kind of book do you think is Pat's favorite? Explain why you think so.

_____

Chapter 11 Review/Test

two hundred five **205**

Spiral Review and Test Prep
Chapters 1–11

① Which shows 47?

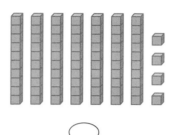 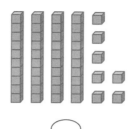 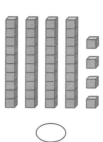 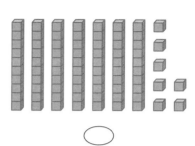
○ ○ ○ ○

② About how many stars are there?

about 5　　　about 10　　　about 50　　　about 100
○　　　　　　○　　　　　　○　　　　　　○

③ 9 + 8 = _____

④ Write the number 65 in words. _____

⑤ Sue started her homework at 3:30. She finished at 5:30. How long did she spend doing homework? Tell how you know.

_____

_____

UNIT 3
CHAPTER 12

Number Relationships and Regrouping

Bushy-Tailed Mathematicians

by Betsy Franco

The squirrels kept their walnuts
in a giant saving jar,
When the jar was full, they counted
all the walnuts saved so far.

First they grouped them all by tens
and then counted one and all,
Though it took them several hours,
all the squirrels had a ball.

When the walnuts were all counted,
then the squirrels used a pen,
wrote the total on a leaf and
put them in the jar again.

Math at Home

Dear Family,

I will learn how to regroup numbers in order to add in Chapter 12. Here are my math words and an activity that we can do together.

Love, _____

My Math Words

regroup:
trade 10 ones for 1 ten

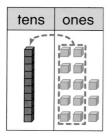

trade 1 ten for 10 ones

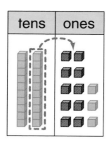

Home Activity

Lay out 16 paper clips or other small objects. Have your child organize 10 clips as a group, and then count the tens and ones that make up 16. Repeat with other numbers between 11 and 29.

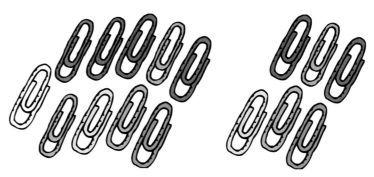

Books to Read

In addition to these library books, look for the Time For Kids math story that your child will bring home at the end of this unit.

- **Double Bubble Trouble!** by Judy Bradbury, Learning Triangle Press, 1998.
- **Five Creatures** by Emily Jenkins, Frances Foster Books, 2001.
- **Time For Kids**

www.mmhmath.com
For Real World Math Activities

Name_____   Explore Regrouping

Learn
Trading 10 ones for 1 ten is called **regrouping**. Use ▬ and ▪ to model 23.

Math Word
regroup

You can show 23 as 23 ones.
You can show 23 as 2 tens 3 ones.

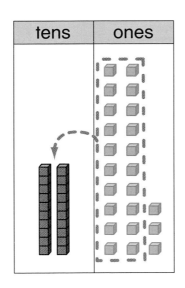

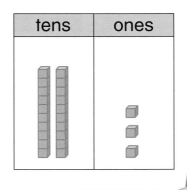

You can regroup 10 ones as 1 ten.

Your Turn
Use ▬, ▪ and a ⬚. Regroup.
Write how many tens and ones.

1) 16 ones = __1__ ten __6__ ones

2) 34 ones = ____ tens ____ ones

3) ✏️ **Write About It!** Explain two different ways to show 17.

Chapter 12 Lesson 1   two hundred nine **209**

Practice Use ▫, ▭ and ▢. Regroup. Write how many tens and ones.

1 ten 13 ones is the same as 2 tens 3 ones.

4) 1 ten 13 ones

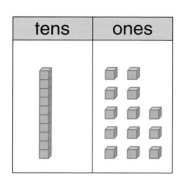

Regroup 10 ones as 1 ten.

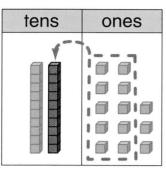

How many tens and ones are there now?

tens	ones

tens	ones
2	3

5) 2 tens 10 ones

tens	ones
___	___

6) 12 ones

tens	ones
___	___

7) 10 ones

tens	ones
___	___

8) 1 ten 18 ones

tens	ones
___	___

9) 15 ones

tens	ones
___	___

10) 1 ten 21 ones

tens	ones
___	___

Spiral Review and Test Prep

11) Which sign goes in the circle?

3 ◯ 5 = 8

= + − ×
◯ ◯ ◯ ◯

12) Julie has 7 books. She gives 3 away. How many books are left?

4   10   11   14
◯   ◯   ◯   ◯

 Math at Home: Your child learned how to regroup 10 ones as 1 ten.
Activity: Have your child tell the number of tens and ones in 17, 24, 33, and 40.

Name _____

Addition with Sums to 20

Learn You can use ▬ and ▫ to model addition.

$9 + 6 = \underline{15}$

Step 1
Show 9 ones.
Show 6 ones.

tens	ones

Step 2
Regroup 10 ones as 1 ten.

tens	ones

Step 3
Write how many tens and ones.

tens	ones

tens	ones
1	5

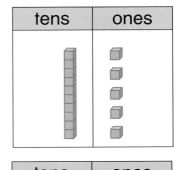

Your Turn Use ▫, ▬ and a ▢.
Write how many tens and ones. Write the sum.

① $8 + 5 = \underline{13}$

tens	ones

tens	ones
1	3

② $7 + 7 = \underline{}$

tens	ones

tens	ones

③ ✎ **Write About It!** Explain how to regroup when you add.

Chapter 12 Lesson 2

Practice Use 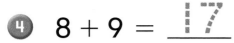 and a ▭.
Write how many tens and ones.
Write the sum.

You can regroup 10 ones as 1 ten.

④ 8 + 9 = 17

tens	ones

tens	ones
1	7

⑤ 8 + 4 = ___

tens	ones

tens	ones

⑥ 7 + 9 = ___

tens	ones

tens	ones

⑦ 9 + 9 = ___

tens	ones

tens	ones

Problem Solving **Mental Math**

> is greater than
< is less than
= is equal to

Compare. Write >, <, or =.

⑧ 5 + 9 ◯ 10

⑨ 2 + 4 ◯ 10

⑩ 5 + 5 ◯ 10

⑪ 6 + 7 ◯ 10

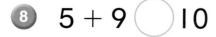

Math at Home: Your child learned how to use a place-value chart to find sums to 20.
Activity: Have your child show you how to use a place-value chart to add 5 + 7.

Name_____

Addition with Greater Numbers

Learn You can use ▭▭▭ and ▫ to model addition.

Add 16 and 25.

Step 1
Show 1 ten 6 ones.
Show 2 tens 5 ones.

Step 2
Combine the ones.
Regroup 10 ones as 1 ten.

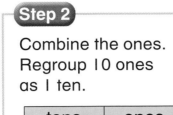
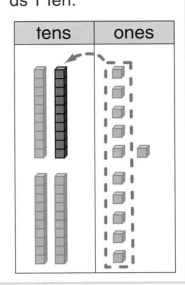

Step 3
How many tens and ones in all?

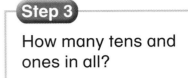
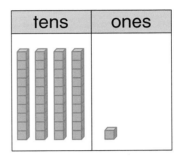

__41__ in all

4 tens 1 one is 41.

Your Turn Use models to add.

Show.	Add. Do you regroup?	How many in all?
① 19 and 24	(yes) no	43
② 31 and 36	yes no	
③ 32 and 18	yes no	

④ ✏️ **Write About It!** When do you regroup?

Chapter 12 Lesson 3

two hundred thirteen **213**

Practice Use ▭, ▪ and [tens|ones] to add.
Regroup when you have 10 or more ones.

5 Add 24 and 18.

Show 2 tens 4 ones. Show 1 ten 8 ones.

Combine the ones. Regroup 10 ones as 1 ten.

Combine the tens. How many tens and ones in all?

tens	ones
4	2

Show.	Add. Do you regroup?	How many in all?
6 17 and 25	yes no	
7 24 and 35	yes no	
8 13 and 27	yes no	

Problem Solving — **Number Sense**

9 How many cookies are there in all? Tell how you found out.

_____ cookies

_____

Math at Home: Your child learned how to use a place-value chart to find the sums of 2-digit numbers.
Activity: Have your child use straws to represent tens and beans to represent ones to add 15 + 17.

Name _____   # Renaming Numbers

Learn You can show numbers in different ways.
Here are three ways to show 23.

2 tens 3 ones   1 ten 13 ones   0 tens 23 ones

Try It Write different ways to show each number.

① 16 __1__ ten __6__ ones   ② 14 ____ ten ____ ones
     __0__ tens __16__ ones   ____ tens ____ ones

③ 19 ____ ten ____ ones   ④ 10 ____ ten ____ ones
     ____ tens ____ ones   ____ tens ____ ones

⑤ 21 ____ tens ____ one   ⑥ 27 ____ tens ____ ones
     ____ ten ____ ones   ____ ten ____ ones
     ____ tens ____ ones   ____ tens ____ ones

⑦ **Write About It!** How do you write a number in different ways?

Practice Write different ways to show each number.

8. 15 — _1_ ten _5_ ones
_0_ tens _15_ ones

9. 11 — ____ ten ____ one
____ tens ____ ones

10. 32 — ____ tens ____ ones
____ tens ____ ones
____ tens ____ ones

11. 36 — ____ tens ____ ones
____ tens ____ ones
____ tens ____ ones

12. 27 — ____ tens ____ ones
____ ten ____ ones
____ tens ____ ones

13. 24 — ____ tens ____ ones
____ ten ____ ones
____ tens ____ ones

Problem Solving / Critical Thinking

14. Circle all the ways to show 28.

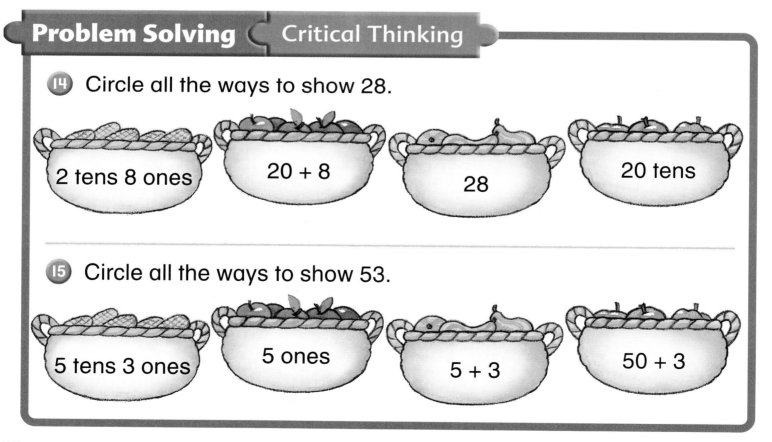

2 tens 8 ones | 20 + 8 | 28 | 20 tens

15. Circle all the ways to show 53.

5 tens 3 ones | 5 ones | 5 + 3 | 50 + 3

Math at Home: Your child learned how to write 2-digit numbers different ways.
Activity: Have your child show three ways to write 29. (2 tens 9 ones, 1 ten 19 ones, 29 ones)

Name _____

Problem Solving Strategy

Use a Pattern

You can use a pattern to solve problems.

Nate picked 8 peaches. He picked 45 apples. How much fruit did he pick in all?

Read

What do I already know? _____ peaches

_____ apples

What do I need to find out? _____

Plan

I know that $8 + 5 = 13$.
I can use a pattern to find $8 + 45$.

Solve

| $8 + 5 =$ | $8 + 15 =$ | $8 + 25 =$ | $8 + 35 =$ | $8 + 45 =$ |

  23

Look Back

I can check my answer using the pattern.

_____

Chapter 12 Lesson 5

Use a pattern to solve.

Draw or write to explain.

1. Kevin picks 7 tomatoes. Marcus picks 35. How many tomatoes do they pick?

   _7_ + _35_ = _42_

   _42_ tomatoes

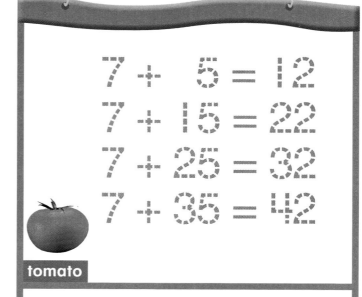

$7 + 5 = 12$
$7 + 15 = 22$
$7 + 25 = 32$
$7 + 35 = 42$

tomato

2. There are 5 big pumpkins and 36 small pumpkins. How many pumpkins are there in all?

   ___ + ___ = ___

   ___ pumpkins

pumpkin

3. There are 6 baskets of corn. The farmer brings 47 more. How many baskets are there in all?

   ___ + ___ = ___

   ___ baskets

corn

Math at Home: Your child added 1-digit and 2-digit numbers using a pattern.
Activity: Have your child show how to use a pattern to solve 7 + 44.

Game Zone

Practice at School ★ Practice at Home

Name _____

Berry Pairs

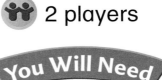 2 players

You Will Need
14 ●

▶ Cover each berry box with a counter.
▶ You and your partner take turns.
▶ Remove 2 counters.
▶ Keep counters if the ways to show numbers match.
▶ Put counters back if they do not.
▶ The player with more counters wins.

Technology Link

Survey and Graphing • Computer

Survey the class. Have your classmates choose which season they like best. Put a tally mark in the table for each classmate.

Favorite Season			
❄ Winter	🌱 Spring	☀ Summer	🍂 Fall

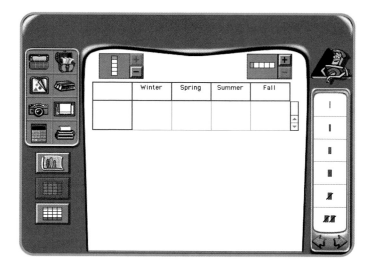

Use [Graphs] to make a bar graph of the survey results.

- Choose a mat to show tally marks.
- Label the columns with the seasons.
- Stamp out the tally marks.
- Click the graph key. Click on Tally.

1. Which season do your classmates like best? _____

2. Do more people like Spring or Fall? _____

3. What if you surveyed all the second grade classes in your school? Which season do you think they would like best? Explain why you think so.

_____

 For more practice use Math Traveler.™

TIME FOR KIDS

On Time

Clocks tell us when to start work or to be at school. School starts at 8:00 A.M. At 8:10 A.M., the kids would be late!

Name _____

These postal workers get the mail out fast. They want to be finished by 4:45 P.M.

On the blank line, write a job you do at home. When do you do this job? Write the time.

_____

___ : ___

Then show the time on this clock.

This officer's shift is over at 11 o'clock.

This nurse starts work at 8 o'clock.

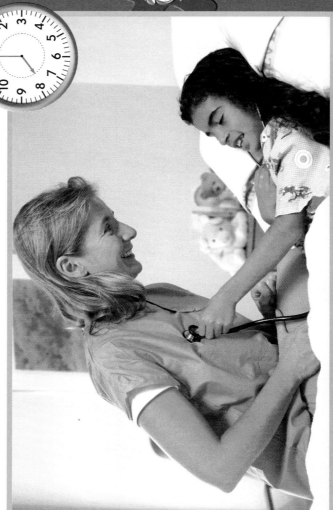

Places open and close at different times.
The library opens at 10 A.M. It closes at 9 P.M.

Name_____

Unit 3
Study Guide and Review

Math Words

Draw lines to match.

1.   — tally marks

2. 60 seconds — one minute

3.  — quarter after 8

Skills and Applications

Time (pages 155–162, 171–180)

Examples

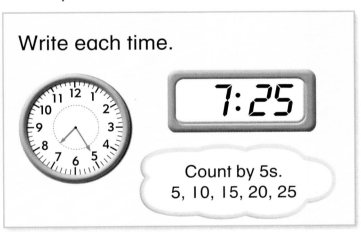

Write each time.
7:25
Count by 5s.
5, 10, 15, 20, 25

4. [clock] __:__

5. [clock] __:__

Use the calendar to answer the question.

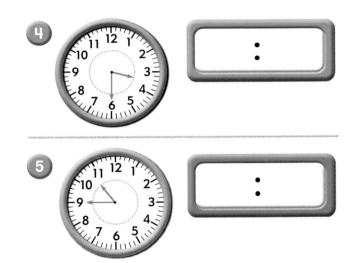

January 24th is Monday.

January

S	M	T	W	T	F	S
						1
2	3	4	5	6	7	8
9	10	11	12	13	14	15
16	17	18	19	20	21	22
23	24	25	26	27	28	29
30	31					

6. What is the date 2 weeks before January 30th?

_____

Skills and Applications

Graphs and Regrouping (pages 189–200, 209–216)

Examples

Use the data from the bar graph to answer questions.

Farm Animals Children Like

	Cow	Horse	Sheep
Votes	7	6	4

0 1 2 3 4 5 6 7 8 9
Number of Votes

7. How many children voted in all? _____

8. How many children voted for sheep? _____

9. Which animal was the favorite?

_____

Regroup. Write how many tens and ones.

1 ten 15 ones

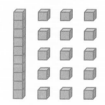

tens	ones
2	5

10. 17 ones

tens	ones

11. 2 tens 14 ones

tens	ones

Problem Solving — Strategy

(pages 181–182, 217–218)

12. Complete the pattern to solve.

20, __30__, 40, 50, __60__, 70

65, 70, ____, 80, 85, ____

Math at Home: Your child practiced time, graphs, and regrouping.
Activity: Have your child use these pages to review time, graphs, and regrouping.

226 two hundred twenty-six

Name_____

Unit 3 Performance Assessment

Puppet Show

Plan a puppet show that has 3 acts.

You Will Need
clock
calendar

- Choose a date.
  Mark the date on the calendar.

  | September | | | | | | |
S	M	T	W	T	F	S
				1	2	3
4	5	6	7	8	9	10
11	12	13	14	15	16	17
18	19	20	21	22	23	24
25	26	27	28	29	30	

- Choose a time.
  Show the time on the clock.

- Complete the poster.
  Include the times for the 3 acts.

PUPPET SHOW!

Date _____

Start Time _____

First Act Starts _____

Second Act Starts _____

Third Act Starts _____

End Time _____

Show Lasts _____

You may want to put this page in your portfolio.

Unit 3
Enrichment

Venn Diagrams

You can use data from a diagram to draw conclusions.

School Clubs

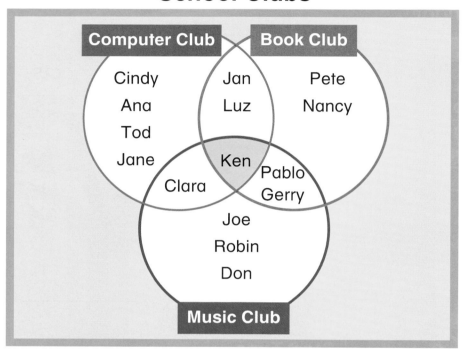

Look! Ken attends all 3 clubs.

Solve.

1. Who attends only the Book Club?

   _____

2. How many children attend the Computer Club? _____ children

3. How many children attend more than 1 club? _____

Show your work.

UNIT 4
CHAPTER 13

2-Digit Addition

Side by Side

Story by Jenny Della Penna • Illustrated by Michael Reid

229

Tim brings 16 paintbrushes.
Kim brings 11.

Who has more paintbrushes?   Tim   Kim

Dan brings 16 cans of paint.
Jan brings the same number of cans of paint.
How many cans of paint did Jan bring?

_____ cans

Ken makes 4 stacks of lumber.
Jen makes 3 stacks of lumber.

How many stacks of lumber in all? _____ stacks

There are 33 ice cold drinks.
Color groups of ten.

There are _____ tens _____ ones.

Math at Home

Dear Family,

I will learn how to add 2-digit numbers in Chapter 13. Here are my math words and an activity that we can do together.

Love, _____

My Math Words

addend :

$$\begin{array}{r} 12 \\ +15 \\ \hline 27 \end{array}$$ ← addend
← addend

sum :

$$\begin{array}{r} 12 \\ +15 \\ \hline 27 \end{array}$$

← sum

regroup :

14 ones = 1 ten 4 ones

Home Activity

Have your child complete the addition wheel for 9.

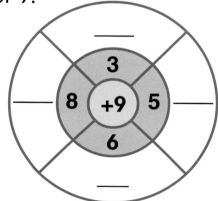

Draw your own addition wheel. Have your child fill in the numbers for other addition facts and complete the wheel.

© Macmillan/McGraw-Hill

Books to Read

Look for these books at your local library and use them to help your child add 2-digit numbers.

- **100 Days of School** by Trudy Harris, Millbrook Press, 1999.
- **Ready, Set, HOP!** by Stuart J. Murphy, HarperCollins, 1996.
- **How Many Teeth?** by Paul Showers, HarperCollins, 1991.

www.mmhmath.com
For Real World Math Activities

Name _____

Mental Math • Add Tens

Math Words

addend

sum

Learn You can use addition facts to help you add tens.

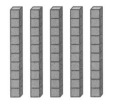 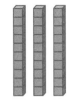

5 + 3 = 8
50 + 30 = 80

5 tens + 3 tens = 8 tens

50 + 30 = 80
↑_addends_↑   ↑sum

Try It Add tens.
You can use ▭.

1. 4 tens + 2 tens = 6 tens

   40 + 20 = 60

2. 6 tens + 1 ten = ___ tens

   60 + 10 = ___

3. 2 tens + 6 tens = ___ tens

   20 + 60 = ___

4. 1 ten + 8 tens = ___ tens

   10 + 80 = ___

5. 3 tens + 4 tens = ___ tens

   30 + 40 = ___

6. 3 tens + 2 tens = ___ tens

   30 + 20 = ___

7. ✏️ **Write About It!** How does knowing 6 + 3 help you find 60 + 30?

Chapter 13 Lesson 1

Practice Add tens.

You can use addition facts to help you add tens.

8) 4 tens   40
   +2 tens   +20
    6 tens   60

9) 5 tens   50
   +4 tens   +40
    ___ tens

10) 30
    +10
     40

11) 50
    +20

12) 60
    +30

13) 40
    +10

14) 20
    +10

15) 20
    +40

16) 50
    +10

17) 20
    +70

18) 30
    +30

19) 10
    +60

20) 20
    +20

21) 60
    +20

Problem Solving — Number Sense

22) Start at 40. Add 10. Add 10 more. Where are you? Tell how you found out.

_____

_____

0 10 20 30 40 50 60 70 80 90 100

Math at Home: Your child used addition facts to add tens.
Activity: Put out 4 dimes and 5 dimes. Have your child tell you how much money there is in all.

Name_____

Memorize your facts. Color to show the strategies you used.

Count on to add.
8 + 3 = 11

Use doubles to add.
6 + 6 = 12
6 + 7 = 13

Make a ten to add.
9 + 5 = 14
8 + 6 = 14

6
+5
__

3
+7
__

7
+7
__

1
+9
__

7 + 5 = ____

9 + 5 = ____

4 + 5 = ____

2 + 9 = ____

8
+8
__

9
+8
__

9
+9
__

7
+8
__

Chapter 13 Extra Practice

two hundred thirty-three **233**

Extra Practice

Use 2 digits from the box. Write a number that makes the comparison true.

1) | 2, 1, 8 |

13 > 12

2) | 4, 5, 1 |

53 < ___

3) | 5, 6, 2 |

64 < ___

4) | 4, 1, 6 |

15 > ___

5) | 3, 7, 5 |

36 > ___

6) | 7, 2, 4 |

73 < ___

7) | 1, 9, 6 |

18 > ___

8) | 8, 3, 5 |

84 < ___

9) | 5, 6, 1 |

62 < ___

10) | 6, 1, 3 |

15 > ___

11) ✏️ **Write About It!** Write the greatest number and the least number you can make with the digits 3, 8, 2.

LOG ON www.mmhmath.com
For more Practice

Math at Home: Your child practiced comparing 2-digit numbers.
Activity: Have your child show you how to compare two 2-digit numbers.

Name _____

Count on Tens and Ones to Add

Learn Use a hundred chart to add.

1	2	3	4	5	6	7	8	9	10
11	12	13	14	15	16	17	18	19	20
21	22	23	24	25	26	27	28	29	30
31	32	33	34	35	36	37	38	39	40
41	42	43	44	45	46	47	48	49	50
51	52	53	54	55	56	57	58	59	60
61	62	63	64	65	66	67	68	69	70
71	72	73	74	75	76	77	78	79	80
81	82	83	84	85	86	87	88	89	90
91	92	93	94	95	96	97	98	99	100

Add 27 + 2. Move across to count on ones.

27 + 2 = 29
2 + 27 = 29

Add 35 + 30. Move down to count on tens.

35 + 30 = 65
30 + 35 = 65

Try It Count on to add. You can use the hundred chart.

1. 36 + 10 = 46
2. 5 + 47 = ___
3. 48 + 3 = ___
4. 14 + 30 = ___
5. 63 + 30 = ___
6. 55 + 30 = ___
7. 29 + 10 = ___
8. 2 + 31 = ___
9. 60 + 18 = ___
10. 47 + 3 = ___

11. **Write About It!** How many tens do you count on to add 48 + 20?

Chapter 13 Lesson 2
two hundred thirty-five **235**

Practice Count on to add.

To count on by tens, keep adding 10 to each number.

12. 30
   +18
   ——
   48

   28 → 38 → 48

13. 49
   + 3

14. 65
   + 2

15. 75
   +10

16. 54
   +30

17.   4
   +23

18. 88
   +10

19.   3
   +64

20. 20
   +29

21. 10
   +37

22. 30
   +16

23. 20
   +47

24. 83
   +10

25. 20
   +33

26. 49
   + 3

27.   2
   +35

Problem Solving Number Sense

Use the number line. Count on to add.

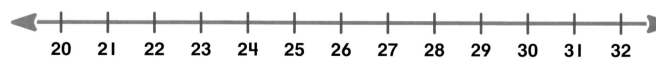

20 21 22 23 24 25 26 27 28 29 30 31 32

28. 23
   + 3
   ——
   26

29. 27
   + 3

30. 29
   + 1

31.   3
   +24

 Math at Home: Your child counted on by tens to add.
Activity: Say a number between 10 and 60. Ask your child to add 30 to the number.

236 two hundred thirty-six

Name_____

Decide When to Regroup

Learn You can use tens and ones models to add. Use the workmat and ▭ and ▫ to add 27 + 5.

Math Word
regroup

Step 1
Show 27 and 5 with ▭ and ▫.

Step 2
Add the ones. If there are 10 or more ones you need to regroup. Replace 10 ones with 1 ten.

Step 3
Write how many tens and ones.

_3_ tens _2_ ones

27 + 5 = _32_

tens	ones

Your Turn Use ▭ and ▫ to add.

Show.	Add the ones. Do you regroup?	How many in all?
① 15 + 8	(yes) no	23
② 23 + 6	yes no	
③ 34 + 7	yes no	

④ **Write About It!** Explain how you know when to regroup ones.

Chapter 13 Lesson 3

two hundred thirty-seven **237**

Practice Use ▬ and ▪ to add.

10 ones equal 1 ten.

Show.	Add the ones. Do you regroup?	How many in all?
5. 43 + 9	yes no	52
6. 26 + 4	yes no	
7. 32 + 6	yes no	
8. 17 + 7	yes no	
9. 29 + 5	yes no	
10. 31 + 8	yes no	
11. 25 + 6	yes no	

Problem Solving — Estimation

 **Show Your Work**

12. If you add 38 and 9 will the sum be less than or greater than 38? Write to explain.

_____

_____

_____

Math at Home: Your child added 2-digit and 1-digit numbers.
Activity: Have your child use straws to represent tens and beans to represent ones to show you how to add 18 + 6.

Name_____

Practice Addition

Learn John and Susie mail letters for their class. John mails 14 letters and Susie mails 16 letters. How many letters did they mail in all?

Add 14 + 16.

tens	ones
¹1	4
+1	6
3	0

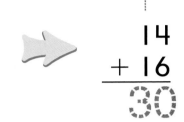

```
  14
+ 16
  30
```

They mailed ___30___ letters in all.

Try It Add. You can use ▭▭▭▭ and ▫ to help.

1)

tens	ones
¹1	9
+2	3
4	2

```
  19
+ 23
  42
```

2)

tens	ones
□ 3	4
+1	1

```
  34
+ 11
```

3)

tens	ones
□ 4	3
+2	7

```
  43
+ 27
```

4)

tens	ones
□ 2	7
+2	5

```
  27
+ 25
```

5) **Write About It!** How do you know when to regroup when you add?

Chapter 13 Lesson 6

two hundred forty-three **243**

Practice Add. You can use ▭ and ▪ to help.

Do you need to regroup?

6.
tens	ones
□	
2	5
+1	4
3	9

25
+14
—
39

7.
tens	ones
□	
1	6
+2	4

16
+24
—

8. 18
   +23

9. 41
   +24

10. 39
   + 6

11. 26
   +25

12. 58
   +10

13. 49
   + 5

14. 13
   +61

15. 28
   +35

16. 14
   +22

17. 17
   +23

Problem Solving — Use Data

Amy sells tickets at the museum.

Museum Tickets Sold
Monday	27
Wednesday	51
Friday	43

18. What is the total number of tickets Amy sold on Monday and Friday?

_____ tickets

19. On which day did Amy sell the most tickets?

_____

Math at Home: Your child added numbers with and without regrouping.
Activity: Ask your child to show you how to add 15 + 12 and how to add 18 + 13.

Name_____

Problem Solving Skill
Reading for Math

Bake Sale

The Kids' Club wants to raise money. So they have a bake sale. They sell 24 bran muffins and 32 corn muffins. They sell 15 blueberry pies, 18 apple pies, and 10 pumpkin pies.

Problem Solving

 Problem and Solution

1. What problem does the Kids' Club have? What do they do to solve this problem?

   _____

   _____

2. How many muffins are there in all? _____ muffins

3. How many pies do they sell in all? _____ pies

4. Do they sell more muffins or pies? _____

Chapter 13 Lesson 7

two hundred forty-five **245**

Making a List

The club has $58. The children need to decide what to buy. They make a list of toys and prices. The club can buy 16 puzzles and 18 games. They can also buy stuffed animals.

Problem and Solution

5. What do the children make a list of? How do they use this list?

   _____

   _____

6. How many puzzles and games do they buy in all?

   _____ puzzles and games

7. The children buy 21 bears, 14 dogs, and 9 cats. How many stuffed animals do they buy in all?

   _____ stuffed animals

Math at Home: Your child identified problems and solutions to answer math questions.
Activity: Ask your child to make a list of his or her toys, or other possessions.

Name _____

Problem Solving Practice

Solve.

1. Max plants 25 🌸.
Jan plants 28 🌸.
How many 🌸 do they plant in all?

____ + ____ = ____

2. Sam picks 16 🍎.
Ann picks 19 .
How many 🍎 do they pick in all?

____ + ____ = ____

3. Sara pulls 38 carrots from the ground. Her mom pulls 27 carrots. How many carrots do they pull in all?

____ + ____ = ____

____ carrots

4. Last week, Nat planted 14 trees. This week, Min planted 18 trees. How many trees did they plant in all?

____ + ____ = ____

____ trees

Write a Story!

5. Use the number sentence. Write an addition story about friends who work together. Find the sum.

35 + 19 = ____

_____

_____

_____

Chapter 13 Problem Solving Practice

two hundred forty-seven **247**

Writing for Math

 Tell your friend how to add 38 and 16. Write the steps in the correct order.

Think

Do I need to regroup?

_____
_____
_____

Solve

I write the steps I use to add 38 + 16.

$$\begin{array}{r} 38 \\ +16 \\ \hline \end{array}$$

_____
_____

Explain

How can I check my answer?

_____
_____
_____

Chapter 13 Review/Test

Name_____

Count on to add.

1) 38 + 3 = _____

2) 19 + 3 = _____

3) 54 + 10 = _____

4) 50 + 42 = _____

Add. You can use ▭ and ▫ to help.

5)
tens	ones
□	
1	7
+ 2	4

6)
tens	ones
□	
2	8
+	5

7)
tens	ones
□	
3	1
+ 2	6

8)
tens	ones
□	
4	7
+ 3	9

Add.

9) 18 + 9

10) 20 + 19

11) 27 + 46

12) 32 + 55

13) 7 + 37

14) 49 + 11

15) 53 + 4

16) 25 + 17

17) 41 + 13

18) 68 + 13

Solve.

19) Tyler reads 18 pages on Monday. He reads 23 pages on Tuesday. How many pages did Tyler read in all?

_____ pages

20) Tom's shelf holds 30 books. He has 15 books on plants. He has 10 books on rocks. Can Tom put all the books on the shelf? Explain.

_____

Spiral Review and Test Prep
Chapters 1–13

Choose the best answer.

1 What time does the clock show?

2:15	2:45	3:15	3:45
○	○	○	○

2 How much money is shown?

17¢	27¢	37¢	45¢
○	○	○	○

3 Jamie starts his homework at 3:30 P.M. He works for 30 minutes. What time does he stop? _____

4 What is the missing number in this pattern?

12, 16, 20, ?, 28, 32

_____

5 Jill has 18 green marbles and 25 blue marbles. How many marbles does Jill have in all? Tell how you found out.

_____ marbles

www.mmhmath.com
For more Review and Test Prep

Practice and Apply
2-Digit Addition

No Room in the Tub

22 crocodiles came for a swim

in my bathtub this morning at eight.

And later 10 more showed up at the door

and said, "We're so sorry we're late."

I'm really delighted my brother invited

the 32 swimmers you see.

But I can't wash my hair or my hands or my face.

There's no room in my bathtub for me!

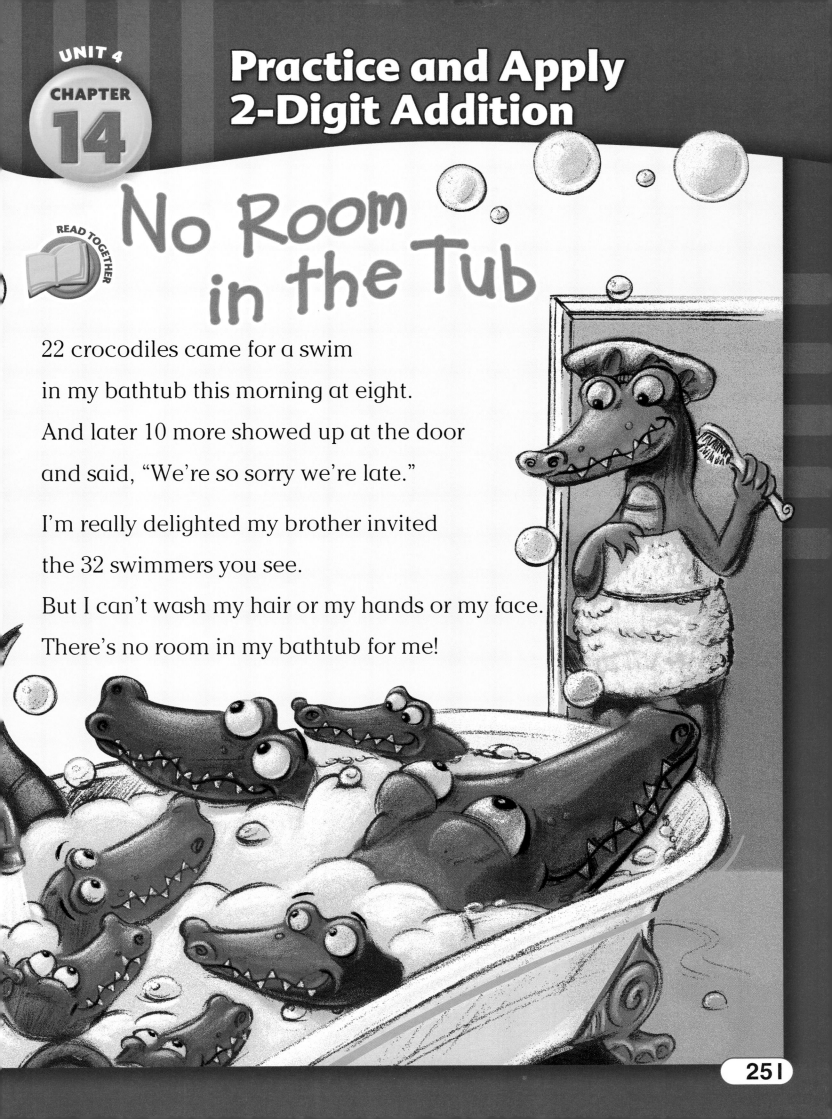

Math at Home

Dear Family,

I will learn more about adding 2-digit numbers in Chapter 14. Here are my math words and an activity that we can do together.

Love, _____

My Math Words

round :

27 rounded to the nearest 10 is 30.

estimate :

number sentence :

26 + 15 = 41

Home Activity

Make a set of number cards. Choose numbers between 0 and 45, such as 18, 25, and so on.

Ask your child to pick 2 cards and add the numbers. Repeat the activity several times.

www.mmhmath.com
For Real World Math Activities

Books to Read

Look for these books at your local library and use them to help your child practice and apply 2-digit addition.

- **Betcha!** By Stuart J. Murphy, HarperCollins, 1997.
- **The Long Wait** by Annie Cobb, The Kane Press, 2000.
- **Mrs. McTats and Her Houseful of Cats** by Alyssa Satin Capucilli, Margaret McElderry Books, 2001.

Name_____   **Check Addition**   ALGEBRA

Learn One way to check addition is by adding the numbers in a different order.

Changing the order of the addends does not change the sum.

Add
1
  35
+28
  63

Check
1
  28
+35
  63

Try It Add. Check by adding in a different order.

1)   47
   +15
    62

    15
   +47
    62

2)   61
   +12

3)   56
    + 9

4)   43
   +29

5)   48
   +32

6)   8
   +76

7) **Write About It!** How can you check your answer when you add?

Chapter 14   Lesson 3   two hundred fifty-seven **257**

Practice Add. Check by adding in a different order.

8)
```
  18        47
+ 47      + 18
  65        65
```

9)
```
  25
+ 30
```

10)
```
   8
 +38
```

11)
```
  73
+  9
```

12)
```
  33
+ 46
```

13)
```
  51
+ 39
```

14)
```
  23
+ 37
```

15)
```
  58
+ 14
```

Spiral Review and Test Prep

Choose the best answer.

16) What time does the clock show?

3:20   4:15   5:15   6:45
○   ○   ○   ○

17) Which has the same sum as 8 + 3?

4+6   7+3   5+7   3+8
○   ○   ○   ○

Math at Home: Your child learned to check addition by changing the order of the addends.
Activity: Ask your child to add 34 + 46. Then have your child show you how to check the answer.

258 two hundred fifty-eight

Name_____

Estimate Sums

Math Words
estimate
reasonable
round

Learn Kim said the sum of 14 + 17 is 31. You can estimate to see if her answer is reasonable.

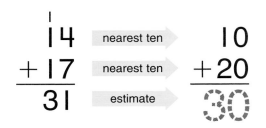

Round numbers to the nearest ten.

14 → nearest ten → 10
+17 → nearest ten → +20
31 → estimate → 30

31 is close to 30.
The answer is reasonable.

Try It Add. Estimate to see if your answer is reasonable. You can use the number line.

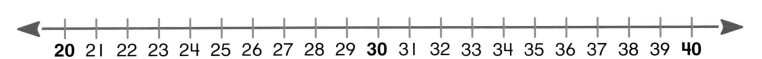

1) 37 → 40
   +29 → +30
   ‾‾‾   ‾‾‾
   66   70

2) 28
   +31

3) 24
   +38

4) 37
   +38

5) 29
   +22

6) 39
   +29

7) **Write About It!** How can you use estimation to see if your answer is reasonable?

Chapter 14 Lesson 4   two hundred fifty-nine **259**

Practice Add. Estimate to see if your answer is reasonable.

Round the numbers to the nearest ten.

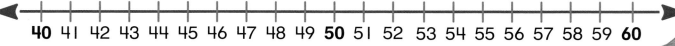

Round 45 to 50.

8) 45 nearest ten → 50
   +42 nearest ten → +40
    87 estimate   → 90

9) 49
   +44

10) 42
   +29

11) 38
   +41

12) 48
   +39

13) 18 + 64 = 82 80

14) 33 + 29 = ___ ___

15) 62 + 26 = ___ ___

16) 25 + 49 = ___ ___

17) 19 + 72 = ___ ___

18) 17 + 48 = ___ ___

Problem Solving — Estimation

19) If Amber sees 28 dog prints and 43 cat prints, will the total number of prints be greater than 100? Is this a reasonable estimate or not? Explain.

_____

Math at Home: Your child estimated sums by finding the nearest ten.
Activity: Ask your child to explain how to estimate 28 + 34.

Name_____

Add Three Numbers

Learn Look for tens or doubles when you add three numbers.

Add 24 + 13 + 36.

Step 1
Add the ones.

```
  2 4
  1 3
+ 3 6
-----
    3
```
10 + 3 = 13

Look for a 10 to help.

Step 2
Add the tens.

```
  2 4
  1 3
+ 3 6
-----
  7 3
```

24 + 13 + 36 = __73__

Try It Look for two numbers in the ones column that make a ten or a double. Circle them. Add.

1. 14
     ⑨
   +3①

    54

2. 32
    42
   +14

3. 25
    15
   +13

4. 62
     8
   +26

5. 23
    31
   +17

6. 35
     9
   +21

7. 13
    23
   +36

8. 47
    22
   +13

9. 51
    12
   +22

10. 20
    25
   +35

11. **Write About It!** How is adding three numbers like adding two numbers?

Chapter 14 Lesson 5

Practice Add.

Add doubles or make a ten first if you can.

12. 13
      3
    +45
    ----
     61   6 + 5 = 11

13. 21
    19
    +30

14. 26
     7
    +46

15. 54
    26
    +15

16. 28
    32
    +23

17. 12
    25
    +21

18. 34
    14
    +17

19. 71
    11
    + 9

20. 43
    10
    +23

21. 27
    36
    +23

22. 34
    13
    +43

23. 3 + 14 + 3 = ___

24. 34 + 18 + 6 = ___

25. 15 + 9 + 15 = ___

26. 12 + 5 + 8 = ___

Problem Solving — Use Data

Use the table.

27. Which cat scored more points? Write to explain how you know.

Points Scored

Cat	Round 1	Round 2	Round 3
Whiskers	18	24	36
Snowball	21	27	27

Math at Home: Your child used mental math to add three numbers.
Activity: Have your child explain how to add 24 + 21 + 16.

Name _____

Problem Solving Strategy

Choose a Method

You can choose a method to solve problems.

Juan takes 16 pictures at the petting zoo. He takes 20 pictures at the pet show. How many pictures does Juan take in all?

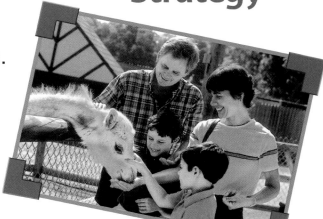

Read

What do I already know? _____ pictures at the petting zoo

_____ pictures at the pet show

What do I need to find? _____

Plan

I can choose a method to solve. I need to decide which way is easier and faster for me to add.

Paper and Pencil   Calculator   Mental Math   Models

Solve

I can use mental math. Count on by tens.

(26, 36)   16 + 20 = 36

Juan takes _____ pictures.

Look Back

How can I check my answer? _____

Read) Plan (Solve) Look Back

Choose a method to solve the problem.

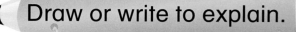

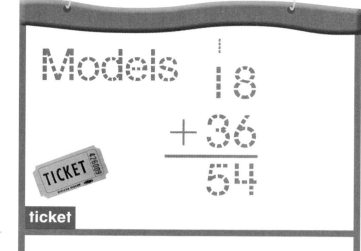

1. Mrs. Lopez buys 18 adult tickets and 36 child tickets for the petting zoo. How many tickets does she buy in all?

   __54__ tickets

2. There are 15 animal balloons. Sara sees 9 more animal balloons. How many animal balloons in all?

   _____ animal balloons

balloon

3. Harry buys 25 bird stickers. He buys 50 fish stickers. How many stickers does Harry buy in all?

   _____ stickers

sticker

4. There are 17 gray goats at the petting zoo. There are 28 black goats and 24 white goats. How many goats are there in all?

   _____ goats

goat

Math at Home: Your child chose a method to solve word problems.
Activity: Ask your child to think of a word problem that would be easy to solve using mental math.

Game Zone

Practice at School ★ Practice at Home

Name _____

Sum It Up!

▶ Take turns. Drop the 2 on the game board.

▶ Write the numbers in a number sentence. Add.

▶ Have your partner check your addition.

▶ After each round the player with the greater sum gets 1 point.

▶ The first player to get 8 points wins.

2 players

You Will Need

2

paper and pencil

Chapter 14 Game Zone

two hundred sixty-five **265**

Technology Link

Addition • Calculator

Use a 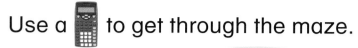 to get through the maze.

If you get stuck, press Clear and start again.

Start at 0.
Find the path that makes the sum of 99.
Color it.

You Will Use

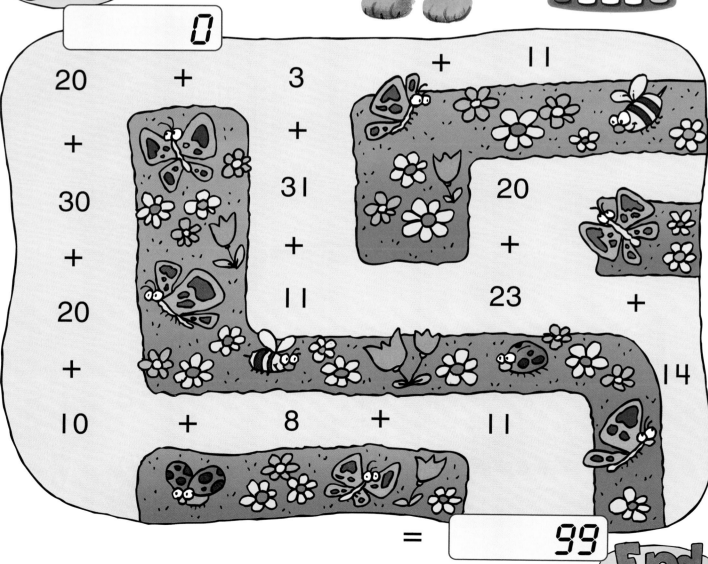

Chapter 14
Review/Test

Name_____

Rewrite the number sentence. Then add.

1. 36 + 18

2. 58 + 9

3. 41 + 37

4. 23 + 17

5. 16 + 45 + 4

6. 3 + 28 + 3

Add. Estimate to see if your answer is reasonable.

7. 34
   +48

8. 38
   +44

Choose mental math or paper and pencil to solve.

9. 38 boys, 40 girls, and 16 adults visit the pet store. How many people visit the pet store?

   _____ people

10. Ashley sees 47 small dog treats. She sees 30 large dog treats. How many dog treats does she see in all?

    _____ dog treats

Chapter 14 Review/Test

two hundred sixty-seven **267**

Spiral Review and Test Prep
Chapters 1–14

Choose the best answer.

1 What number comes next?

2, 4, 8, 16, ____

- 18
- 20
- 24
- 32

2 Start at 38. Add 20. Where are you?

- 28
- 48
- 58
- 68

3 It takes Rebecca about 10 seconds to tie a bow. How many bows can she tie in 1 minute?

- 6
- 10
- 16
- 60

Answer these questions.

4 What is 30¢ more than 1 quarter and 2 pennies?

____¢

5 It is 6:15 A.M. What time will it be in 20 minutes?

`6:15`

___ : ___ A.M.

6 Write 2 different ways to show a sum of 60.

___ + ___ = 60   ___ + ___

LOG ON www.mmhmath.com
For more Review and Test Prep

UNIT 4 CHAPTER 15

2-Digit Subtraction

Tea Party

Molly and I had a party

In hats—red for her, green for me.

A platter of eighty-six peanuts was served

Along with a big pot of tea.

Molly ate eighty-four peanuts,

Carefully leaving me two.

Did I say that my hungry friend Molly

Is an elephant who lives at the zoo?

Math at Home

Dear Family,

I will learn how to subtract 2-digit numbers in Chapter 15. Here are my math words and an activity that we can do together.

Love, _____

My Math Words

difference :

$$\begin{array}{r}36\\-12\\\hline 24\end{array}$$ ← difference

regroup :

1 ten 3 ones = 13 ones

Home Activity

Have your child complete the subtraction square. Subtract across all rows. Subtract down all columns.

Use different subtraction facts to make your own subtraction square.

Books to Read

Look for these books at your local library and use them to help your child subtract 2-digit numbers.

- **Bunny Money** by Rosemary Wells, Viking, 1997.
- **The Grapes of Math** by Greg Tang, Scholastic, 2001.

www.mmhmath.com
For Real World Math Activities

Name_____

Mental Math • Subtract Tens

Learn You can use subtraction facts to help you subtract tens.

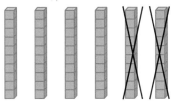

$6 - 2 = 4$
So, $60 - 20 = 40$

6 tens − 2 tens = __4__ tens

60 − 20 = __40__

Try It Subtract tens.

1. 8 tens − 5 tens = __3__ tens   80 − 50 = __30__

2. 5 tens − 4 tens = ____ ten   50 − 40 = ____

3. 6 tens − 4 tens = ____ tens   60 − 40 = ____

4. 8 tens − 6 tens = ____ tens   80 − 60 = ____

5. 4 tens − 3 tens = ____ ten   40 − 30 = ____

6. 9 tens − 6 tens = ____ tens   90 − 60 = ____

7. 7 tens − 2 tens = ____ tens   70 − 20 = ____

8. **Write About It!** How does knowing $7 - 2$ help you find $70 - 20$?

Chapter 15 Lesson 1

Practice Subtract tens.

You can use subtraction facts to help you subtract tens.

⑨ 5 tens − 3 tens = __2__ tens   50 − 30 = __20__

⑩ 7 tens − 4 tens = _____ tens   70 − 40 = _____

⑪ 60 − 50	⑫ 90 − 20	⑬ 80 − 20	⑭ 30 − 10	⑮ 70 − 50
⑯ 80 − 70	⑰ 50 − 20	⑱ 60 − 30	⑲ 90 − 40	⑳ 80 − 40
㉑ 60 − 20	㉒ 70 − 10	㉓ 90 − 70	㉔ 80 − 50	㉕ 50 − 10

Problem Solving **Number Sense**

Show Your Work

㉖ Jane has 8 dimes in her purse. She gives Jack 3 dimes. How much money does Jane have left?

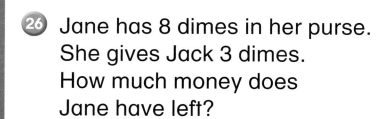

_____ ¢ − _____ ¢ = _____ ¢

Math at Home: Your child used subtraction facts to subtract tens.
Activity: Put 5 dimes on a table. Then cover 3 dimes. Have your child tell you how much money is left.

Name_____

Add or subtract. Fill in the missing numbers.

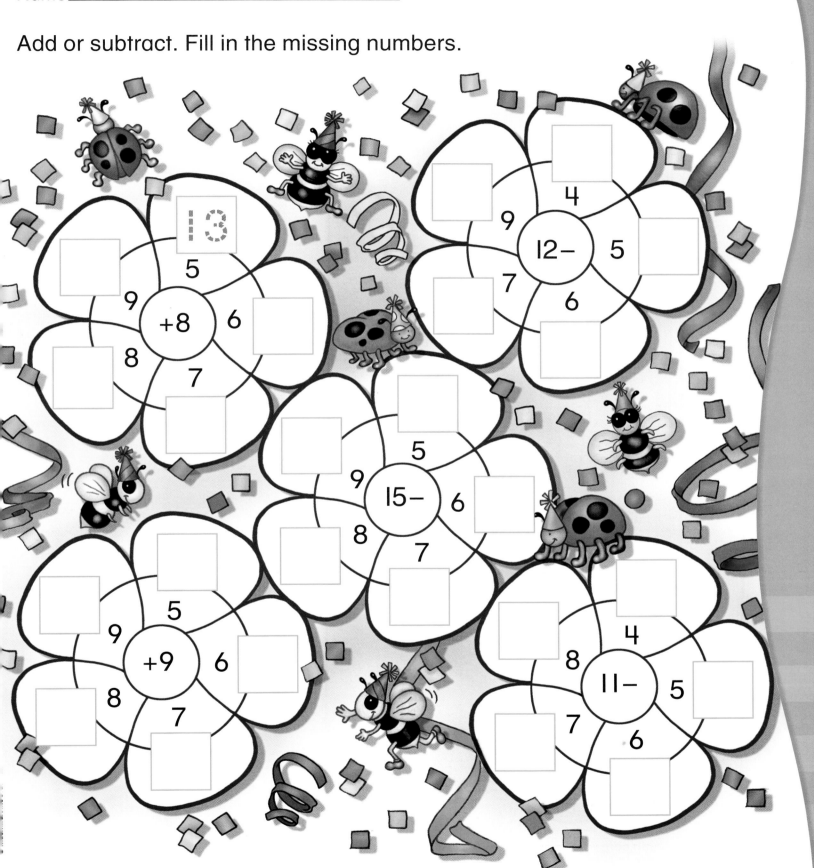

✎ Write **About It!** Write three subtraction facts that have a difference of 8.

Extra Practice

Count the money.
Write the total amount.

Use $ and . to write dollars and cents.

1.

   Total: $1.51

2.

   Total: $ __.__

3.

   Total: $ __.__

4.

   Total: $ __.__

5.

   Total: $ __.__

www.mmhmath.com
For more Practice

 Math at Home: Your child practiced counting dollars and cents.
Activity: Show your child some dollars and coins. Have your child count the total amount of money.

Name_____

Mental Math • Count Back Tens and Ones to Subtract

Learn Use a hundred chart.
Subtract 28 − 3. Subtract 68 − 20.

1	2	3	4	5	6	7	8	9	10
11	12	13	14	15	16	17	18	19	20
21	22	23	24	25	26	27	28	29	30
31	32	33	34	35	36	37	38	39	40
41	42	43	44	45	46	47	48	49	50
51	52	53	54	55	56	57	58	59	60
61	62	63	64	65	66	67	68	69	70
71	72	73	74	75	76	77	78	79	80
81	82	83	84	85	86	87	88	89	90
91	92	93	94	95	96	97	98	99	100

Move left to count back by ones.

Move up to count back by tens.

28 − 3 = **25**   68 − 20 = ____

Try It Count back to subtract.

① 47 − 10 = **37**   ② 38 − 20 = ____   ③ 76 − 3 = ____

④ 82 − 30 = ____   ⑤ 92 − 10 = ____   ⑥ 59 − 30 = ____

⑦ 65 − 20 = ____   ⑧ 77 − 3 = ____   ⑨ 41 − 20 = ____

⑩ 52 − 2 = ____   ⑪ 28 − 20 = ____   ⑫ 36 − 30 = ____

⑬ **Write About It!** How many tens do you count back to subtract 65 − 30?

Practice Count back by tens or ones to subtract.

To subtract by tens, count back by tens.

14. 72 − 30 = 42   62 52 42
15. 93 − 30
16. 84 − 20
17. 36 − 1

18. 80 − 20
19. 18 − 3
20. 54 − 30
21. 95 − 2
22. 27 − 10

23. 39 − 1
24. 61 − 20
25. 70 − 30
26. 22 − 2
27. 43 − 20

28. 57 − 30
29. 35 − 10
30. 49 − 1
31. 26 − 20
32. 45 − 20

Algebra • Input/Output Tables

Find the missing numbers. Follow the rule.

33. **Rule: Subtract 1**

In	Out
36	35
37	36
38	

34. **Rule: Subtract 2**

In	Out
36	
37	
38	

35. **Rule: Subtract 20**

In	Out
36	
37	
38	

Math at Home: Your child counted back by tens to subtract.
Activity: Say a number between 40 and 90. Ask your child to subtract 20 from the number.

Name_____

Decide When to Regroup

Learn You can use tens and ones models to subtract. Use the workmat and and ▫ to subtract 24 − 8.

Step 1
Show 24.

tens	ones

2 tens 4 ones

Step 2
Subtract the ones. There are not enough ones to subtract. Regroup.

tens	ones

1 ten 14 ones

Step 3
Subtract 8 ones.

tens	ones

1 ten 6 ones

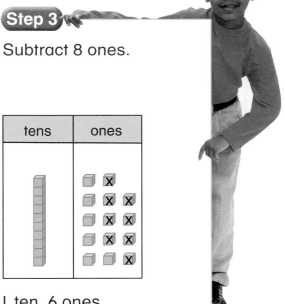

24 − 8 = 16

Your Turn Use  and ▫ to subtract.

Show the first number.	Subtract the ones. Do you need to regroup?	How many are left?
① 31 − 4	(yes) no	27
② 27 − 5	yes no	
③ 42 − 6	yes no	

④ ✏️ **Write About It!** Explain how you know when to regroup to subtract.

Chapter 15 Lesson 3

Practice Use ▭ and ▫ to subtract.

If there are not enough ones to subtract, regroup 1 ten as 10 ones.

Show the first number.	Subtract the ones. Do you need to regroup?	How many are left?
5. 35 − 9	yes no	26
6. 40 − 8	yes no	
7. 56 − 5	yes no	
8. 21 − 6	yes no	
9. 33 − 4	yes no	
10. 43 − 3	yes no	
11. 25 − 7	yes no	

Problem Solving Critical Thinking

12. If you subtract 9 from 27 will the difference be less than or greater than 27? Write to explain.

_____

_____

Math at Home: Your child subtracted two numbers, deciding each time if regrouping was needed.
Activity: Have your child use 2 bundles of 10 straws and 3 single straws to represent 23. Ask your child to show you how to subtract 23 − 5.

Name_____

Subtract a 1-Digit Number from a 2-Digit Number

Learn Subtract 43 − 7.

Step 1

Show 43.
Are there enough ones to subtract 7 ones?

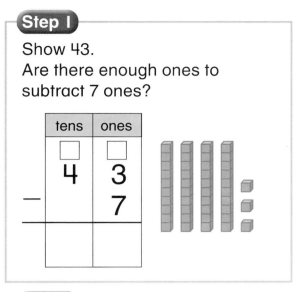

Step 2

There are not enough ones to subtract.
Regroup 1 ten as 10 ones.
Now there are 13 ones.

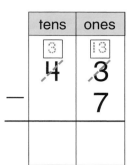
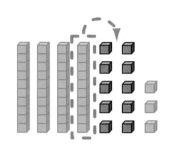

Step 3

Subtract the ones.

Step 4

Subtract the tens.

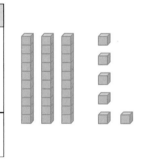

43 − 7 = 36

Try It Subtract. You can use ▭ and ▫ to help.

1.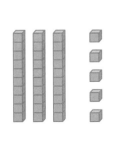

2.
tens	ones
2	3
−	5

3. **Write About It!** How did you show you regrouped 1 ten in exercise 2?

Chapter 15   Lesson 4

Practice Subtract. You can use 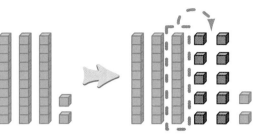 and ▫ to help.

If there are not enough ones to subtract you need to regroup.

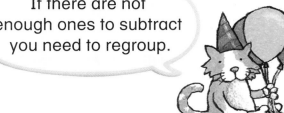

4.
tens	ones
2	12
3̷	2̷
−	4
2	8

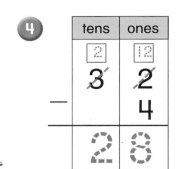

5.
tens	ones
☐	☐
4	1
−	3

6.
tens	ones
☐	☐
2	2
−	9

7.
tens	ones
☐	☐
5	4
−	4

8.
tens	ones
☐	☐
6	0
−	4

9.
tens	ones
☐	☐
3	1
−	6

10.
tens	ones
☐	☐
2	5
−	8

Spiral Review and Test Prep

Choose the best answer.

11. Lena's birthday party started at 4:30 P.M. It ended at 6:30 P.M. How long did Lena's birthday party last?

    two hours ○   three hours ○   four hours ○

12. My number has 2 digits. It is odd. There is a 3 in the tens place. It is more than 31, and less than 35.

    What is the number?
    31 ○   33 ○   34 ○

 Math at Home: Your child subtracted numbers with regrouping.
Activity: Have your child show you how to subtract 6 from 43.

Name _____

Subtract 2-Digit Numbers

Learn Subtract 53 − 17.

Step 1

Show 53.
Are there enough ones to subtract 7 ones?

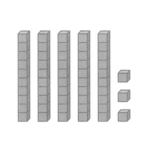

Step 2

There are not enough ones to subtract.
Regroup 1 ten as 10 ones.
Now there are 13 ones.

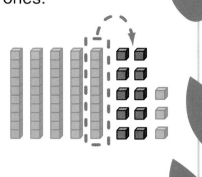

Step 3

Subtract the ones.

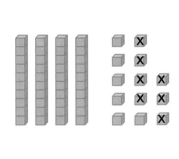

Step 4

Subtract the tens.

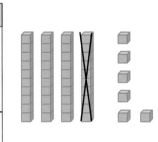

53 − 17 = __36__

Try It Subtract. You can use ▭▭▭ and ▫ to help.

1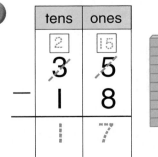

2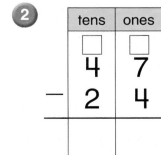

3 ✏️ **Write About It!** How is subtracting 42 − 26 different from subtracting 42 − 6?

Chapter 15   Lesson 5   two hundred eighty-one **281**

Practice

Subtract. You can use ▭ and ◾.

If there are not enough ones to subtract you need to regroup.

4.
tens	ones
2	12
3̷	2̷
− 2	4
	8

5.
tens	ones
□	□
8	1
− 2	3

6.
tens	ones
□	□
6	4
− 3	4

7.
tens	ones
□	□
3	1
− 2	6

8.
tens	ones
□	□
9	5
− 6	8

9.
tens	ones
□	□
5	3
− 4	7

10.
tens	ones
□	□
7	2
− 2	9

Problem Solving — Mental Math

11. Chris counts 18 . He counts 7 . How many more than does he count?

_____

12. Danielle makes 6 🎀. She makes 19 🎀. How many more 🎀 than 🎀 bows does she make?

_____ 🎀

Math at Home: Your child subtracted numbers with regrouping.
Activity: Have your child show you how to subtract 26 from 43.

Name_____

Practice Subtraction

Learn Brian finds 16 shells at the beach. Paige finds 30 shells. How many more shells does Paige find than Brian?

Show 30.
Look at the ones.
Regroup if you need to.

You can subtract to find how many more.

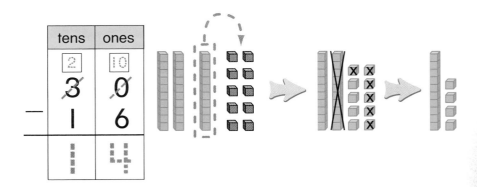

Paige finds ___14___ more shells than Brian.

Try It Subtract. You can use ▬ and ▪ to help.

1.
tens	ones
4⃠5	11⃠1
3	8
1	3

2.
tens	ones
☐	☐
4	7
2	4

3.
tens	ones
☐	☐
9	5
1	8

4.
tens	ones
☐	☐
7	0
1	4

5. **Write About It!** Do you always need to regroup when you subtract? Write to explain.

Chapter 15 Lesson 6

Practice Subtract. You can use ▬ and ▪ to help.
Color answers greater than 50 🖍.
Color answers less than 50 🖍.

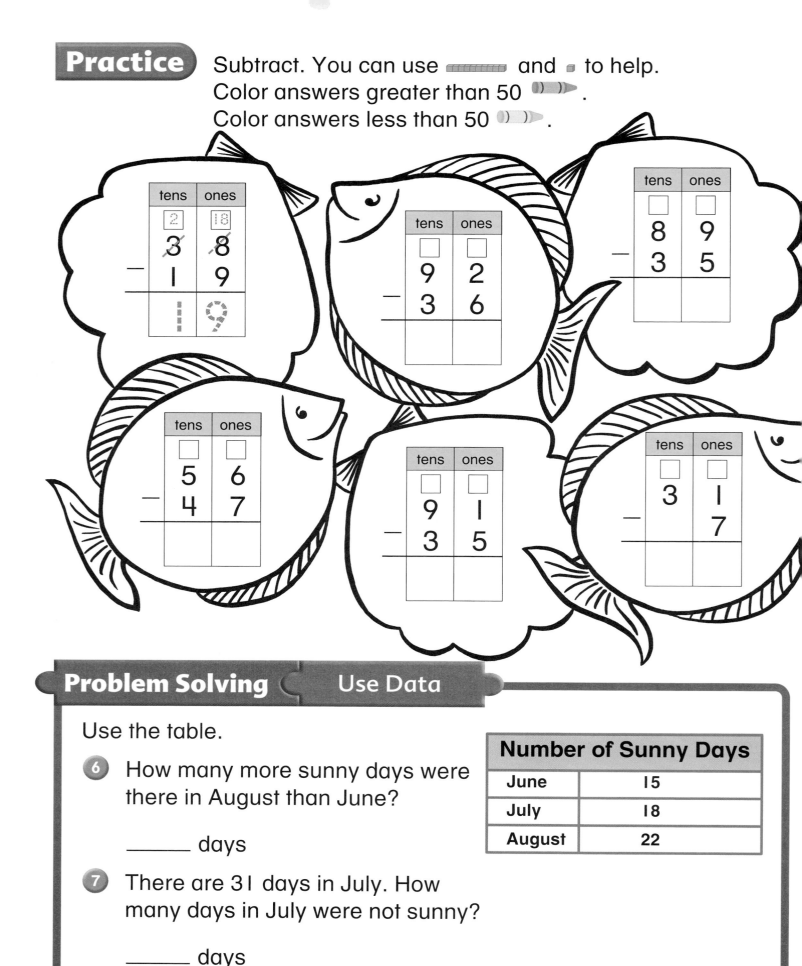

Problem Solving — Use Data

Use the table.

6. How many more sunny days were there in August than June?

   _____ days

Number of Sunny Days	
June	15
July	18
August	22

7. There are 31 days in July. How many days in July were not sunny?

   _____ days

Math at Home: Your child subtracted numbers with and without regrouping.
Activity: Ask your child to show you how to subtract 24 – 15 and how to subtract 36 – 11.

Name_____

Problem Solving Skill
Reading for Math

Apple Picking

Tina's family drove to a farm to pick fruit. First they picked 41 red apples. Next they ate 6 of them. Then they picked 22 green apples. Dad said, "I think we have enough apples!"

 Sequence of Events

1. Did they pick green apples before or after they picked red apples?

   _____

2. Did they pick a greater number of apples before or after they ate?

   _____

3. How many red apples are they taking home? Write a subtraction sentence.

   _____   _____ red apples

Chapter 15   Lesson 7

Picking Vegetables

Tina's family walked to the vegetable garden. First they picked 33 tomatoes. Next they picked 15 green peppers. Then they picked 26 carrots. At last they put all the food in the car and drove home.

Sequence of Events

4. What did the family pick first, tomatoes or green peppers?

   _____

5. Which vegetable did the family pick before the carrots?

   _____

6. How many more of the first vegetable did they pick than the last vegetable? Write a subtraction sentence.

   _____

Math at Home: Your child followed a sequence of events to answer questions.
Activity: Ask your child to list, in order, the vegetables that were picked in the story. Then ask: *How many more carrots than peppers were picked?*

Name _____

Problem Solving Practice

Solve.

 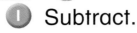 **Write a Story!**

1) Subtract. Then draw a picture to show the subtraction.

24 − 9 = _____

2) There are 28 pumpkins for sale. 7 pumpkins are sold. How many pumpkins are left?

28 − 7 = _____ pumpkins

3) There are 21 birds in a tree. 15 birds fly away. How many birds are left?

21 − 15 = _____ birds

4) Marta picks 34 apples. She gives 18 of them to her friends. How many apples does Marta have left?

34 − 18 = _____ apples

5) The children build 26 snowmen. Then 9 snowmen melt. How many snowmen are left?

26 − 9 = _____ snowmen

Chapter 15 Problem Solving Practice

two hundred eighty-seven **287**

Writing for Math

 Write a subtraction story. Use the subtraction sentence.

$34 - 16 =$ _____

Think

What numbers do I need to use in my story? _____

How can I make my story a subtraction story?

_____

_____

_____

Solve

Next I write my subtraction story and solve the subtraction sentence.

_____

_____

$34 - 16 =$ _____

Explain

This is how my story matches the subtraction sentence.

_____

_____

_____

Chapter 15
Review/Test

Name _____

Count back to subtract.

1) 72 − 20 = ____

2) 97 − 30 = ____

Subtract.

3)
tens	ones
4	7
− 1	5

4)
tens	ones
2	8
−	5

5)
tens	ones
5	4
−	5

6)
tens	ones
2	1
− 1	7

7)
tens	ones
6	2
− 2	5

8)
tens	ones
7	0
− 4	1

Solve.

9) There are 24 children skating.
Then 8 children go home.
How many children are still skating?

_____ children

10) There are 31 boys and 12 girls sledding.
How many more boys than girls
are sledding?

_____ more boys

Chapter 15 Review/Test

two hundred eighty-nine **289**

Spiral Review and Test Prep
Chapters 1–15

Choose the best answer.

1. Start at 38. Add 10. Add 4 more. Where are you?

   48 ○   52 ○   58 ○   62 ○

2. 26 − 15 = ☐

   5 ○   11 ○   41 ○   53 ○

Write the answer.

3. Emily has 23 stickers.
   She gives her sister 18 stickers.
   How many stickers does Emily have left? _____

4. Jim starts his homework at 4:15 P.M.
   He finishes in a half hour.
   What time does Jim finish his homework?

5. What comes next in the pattern?

   23, 26, 29, 32, 35, 38, _____

6. Find the number to complete the number sentence.

   ___ + 10 = 57

   Tell how you know.

www.mmhmath.com
For more Review and Test Prep

Practice and Apply 2-Digit Subtraction

UNIT 4 CHAPTER 16

Pancake Song

Sung to the tune of "On Top of Old Smoky"

Dad's cooking the breakfast—
It's Dad's Pancake Day!
When Dad's making pancakes,
Stay out of the way.

He juggles and flips them—
Some fly through the door.
Some land on the stove top.
Some fall on the floor.

Dad started with fifty.
He dropped twenty-one.
Enough are left over
To feed everyone.

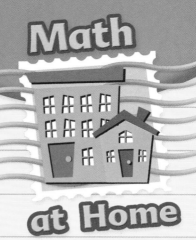

Math at Home

Dear Family,

I will learn more about subtracting 2-digit numbers in Chapter 16. Here is my math word and an activity that we can do together.

Love, _____

My Math Word

estimate :

Home Activity

Make twelve 2-digit number cards. Choose numbers between 10 and 90, such as 12, 25, 34, and 76.

Ask your child to choose 2 cards. Have your child use paper and pencil to subtract the lesser number from the greater number. Repeat the activity several times.

Books to Read

In addition to these library books, look for the Time For Kids math story that your child will bring home at the end of this unit.

- **Alexander, Who Used to Be Rich Last Sunday** by Judith Viorst, Atheneum Books, 1978.
- **The Relatives Came** by Cynthia Rylant, Bradbury Press, 1985.
- **Time For Kids**

www.mmhmath.com
For Real World Math Activities

292 two hundred ninety-two

Name_____

Estimate Differences

Learn Anthony says the difference of 23 − 14 is 9. You can estimate to see if his answer is reasonable.

Math Word
estimate

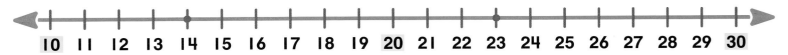

Find the nearest 10 for each number.

```
  23   nearest ten →   20
 −14   nearest ten →  −10
 ───   estimate    →   ──
   9                   10
```

9 is close to 10.
His answer is reasonable.

Which ten is 9 closest to?

Try It Subtract. Use a number line to estimate the difference.

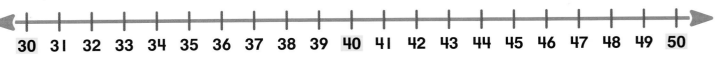

① 49 50	② 43	③ 41
−31 −30	−34 −	−34 −
── ──		
18 20		

| ④ 38 | ⑤ 41 | ⑥ 47 |
| −31 − | −32 − | −36 − |

⑦ ✏️ **Write About It!** How do you estimate to see if your answer is reasonable?

Chapter 16   Lesson 4   two hundred ninety-nine **299**

Practice Subtract. Estimate to see if your answer is reasonable.

Rewrite the numbers to the nearest ten.

8) 55 nearest ten → 60
  −31 nearest ten → −30
  ‾‾24 estimate → ‾‾30

Round 55 up to 60.

9) 46   —
  −27   ___

10) 36   —
   − 8   ___

11) 77   —
   −48   ___

12) 25   —
   −19   ___

13) 94   —
   −45   ___

14) 59   —
   −21   ___

15) 88   —
   −39   ___

Problem Solving — Estimation

Show Your Work

16) There are 56 boats at the dock. 18 boats sail away. About how many boats are left at the dock? Circle the answer that shows about how many.

more than 60   less than 60

Math at Home: Your child estimated differences by finding the nearest ten.
Activity: Ask your child to explain how to estimate 68 − 41.

300 three hundred

Name_____

Mental Math • Strategies

Learn You can look for ways to add and subtract without using paper and pencil.

This is how Lee added 47 + 25.

$$47 + 25 \rightarrow 50 + 25 = 75 \rightarrow 75 - 3 = 72$$

47 + 25 = __72__

I can add 50 and 25. Since 50 is 3 more than 47, I need to subtract 3 from the sum.

I can subtract 40 from 52. Since 38 is 2 less than 40, I need to add 2 to the difference.

This is how Jan subtracted 52 − 38.

$$52 - 38 \rightarrow 52 - 40 = 12 \rightarrow 12 + 2 = 14$$

52 − 38 = __14__

Try It Add or subtract without using paper and pencil.

1. 36
   +29
   ___
   65

   36 + 30 = 66, 66 − 1 = 65

2. 18
   +45
   ___

3. 29
   +31
   ___

4. 62
   −29
   ___

5. 58
   −40
   ___

6. 37
   −19
   ___

7. 42
   −27
   ___

8. **Write About It!** How did you do the subtraction in exercise 4?

Chapter 16 Lesson 5   three hundred one **301**

Practice Add or subtract without using paper and pencil.

9.
$$64 - 9 \rightarrow 64 - 10 = 54 \rightarrow 54 + 1 = 55$$

9 is close to 10.
64 − 10 = 54.
54 + 1 = 55.

10. 54 + 27
11. 23 + 39
12. 27 + 42
13. 28 + 19
14. 37 + 26

15. 57 − 28
16. 71 − 9
17. 41 − 18
18. 63 − 19
19. 38 − 19

 **Algebra • Missing Number**

Find the missing number.

20. 24 + ☐ = 54
21. 32 + ☐ = 40
22. 47 − ☐ = 27
23. 68 − ☐ = 60
24. 33 + ☐ = 63
25. 22 − ☐ = 20

 Math at Home: Your child used mental math to add and subtract.
Activity: Have your child explain how to subtract 53 − 29 using mental math.

Problem Solving Strategy

Choose the Operation

Sometimes you need to choose the operation to solve a problem.

22 people are at the top of the hill. 3 people sled down. How many people are left at the top of the hill?

Read

What do I already know? _____ people are at the top of the hill

_____ people sled down

What do I need to find? _____

Plan

Do I add or subtract to find the answer? I need to find how many are left. I will __subtract__.

Solve

I can carry out my plan.

$$\begin{array}{r} 22 \\ -3 \\ \hline \end{array}$$

_____ people are left.

Look Back

How can I check my answer?

_____

Chapter 16 Lesson 6

three hundred three **303**

Circle add or subtract. Solve.

Draw or write to explain.

1. Kate collects 18 orange leaves and 15 yellow leaves. How many leaves does she collect in all?

   (add)   subtract

   18 (+) 15 = 33

   33 leaves

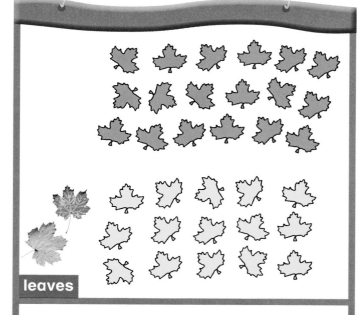
leaves

2. The children make 34 paper snowflakes. 18 snowflakes are white. How many snowflakes are not white?

   add   subtract

   ____ ◯ ____ = ____

   ____ snowflakes are not white

snowflakes

3. Alex picks 32 apples. Carl picks 28 apples. How many apples did the boys pick?

   add   subtract

   ____ ◯ ____ = ____

   ____ apples

apple

Math at Home: Your child learned to choose the operation to solve problems.
Activity: Ask your child to tell a math story that can be solved by subtracting.

Name_____

Game Zone

Practice at School ★ Practice at Home

👥 2 players

The Least Difference

▶ You want to make two numbers that will give you the least difference when you subtract.
▶ Take turns.
▶ Each player spins 4 digits.
▶ Use the 4 digits to make two numbers.
▶ Subtract the smaller number from the greater number.
▶ The player with the least difference wins.

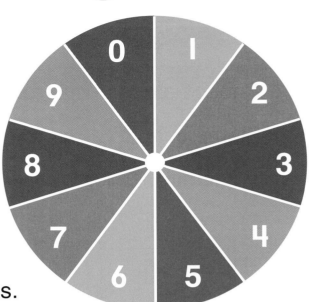

Round 1
Player 1   Player 2

Round 2
Player 1   Player 2

Round 3
Player 1   Player 2

Round 4
Player 1   Player 2

Chapter 16   Game Zone

three hundred five   **305**

Technology Link

Use Place Value to Subtract • Computer

Use to help you subtract.

53 − 16 = _____

- Choose a mat to subtract.
- Stamp out 53. Click on −.
- Trade one 10 down.
- Click on 16.
- Find the difference.

53 − 16 = 37

Subtract. You can use .

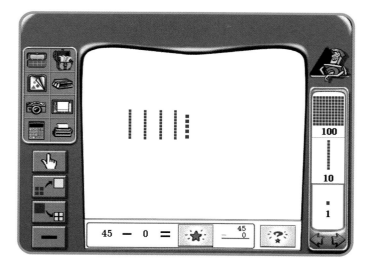

① 45 − 24 = _____

② 45 − 18 = _____

③ 45 − 25 = _____

④ 45 − 36 = _____

 For more practice use Math Traveler.™

Chapter 16
Review/Test

Name_____

Subtract. Check by adding.

① 42　　　　+___
　 −23

② 34　　　　+___
　 −26

Subtract. Estimate to see if your answer is reasonable. You can use the number line.

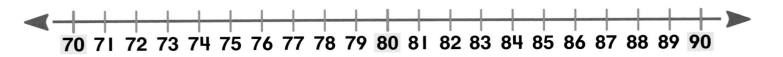

③ 87　　　−___
　 −72

④ 82　　　−___
　 −74

Solve this problem.

⑤ Daniel and Marta play a game. Daniel scores 36. Marta scores 25. Write a number sentence to show how many more points Daniel scored than Marta.

____ ◯ ____ = ____

Chapter 16　Review/Test　　　three hundred seven **307**

Spiral Review and Test Prep
Chapters 1–16

Choose the best answer.

1 What time does the clock show?

9:20 ◯   6:45 ◯   5:45 ◯   5:15 ◯

2 How much money is shown?

20¢ ◯   55¢ ◯   60¢ ◯   80¢ ◯

3 23 + 15 + 47 = _____

4 Double me and you get 24. What number am I? _____

5 Carl has 34 baseball cards. He gives his sister 20 baseball cards. How many baseball cards does Carl have left? Tell how you know.

_____ baseball cards

www.mmhmath.com
For more Review and Test Prep

TIME FOR KIDS

Name _____

Chimpanzees eat about 80 different kinds of food. Koalas eat only 1 kind of leaf.

How many more kinds of food does a chimpanzee eat than a koala?

____ – ____ = ____

About how many kinds of food do you like to eat?

_____

TIME FOR KIDS

What's to Eat?

All animals eat. Pandas mostly eat bamboo plants. They eat about 40 pounds of bamboo each day.

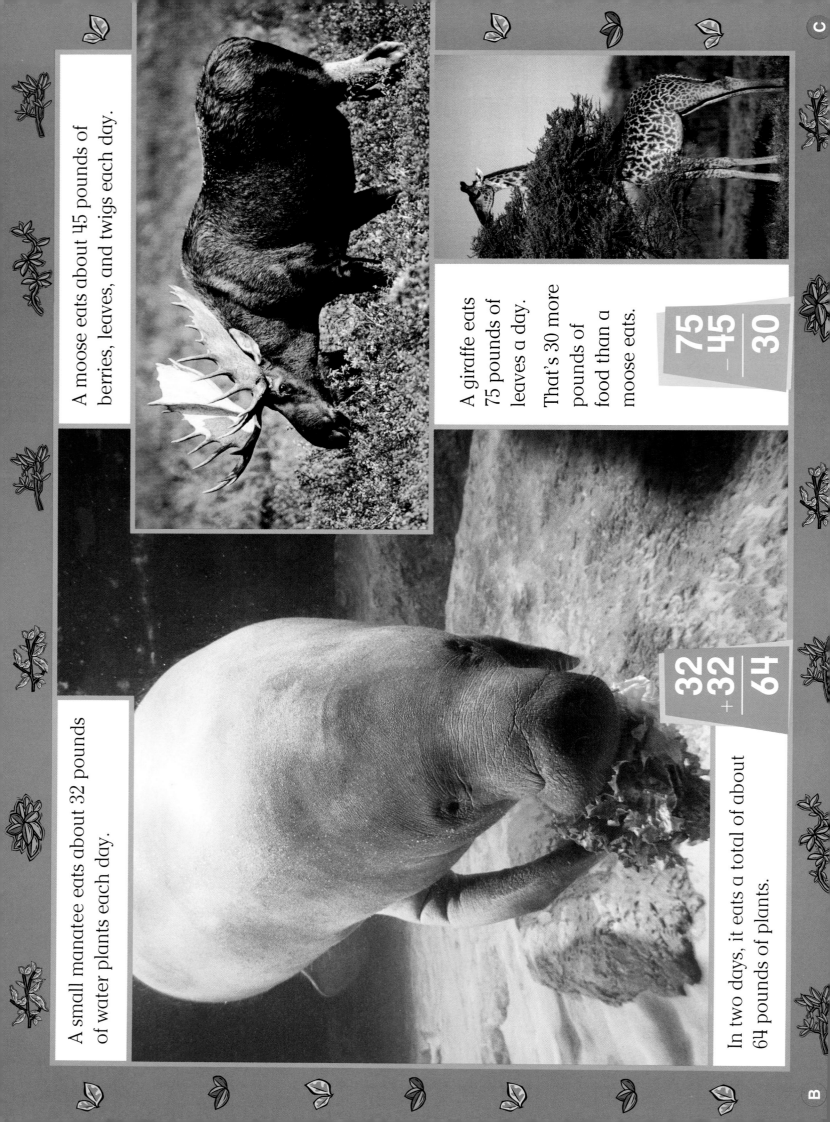

A moose eats about 45 pounds of berries, leaves, and twigs each day.

A giraffe eats 75 pounds of leaves a day.

That's 30 more pounds of food than a moose eats.

$$\begin{array}{r} 75 \\ -45 \\ \hline 30 \end{array}$$

A small manatee eats about 32 pounds of water plants each day.

In two days, it eats a total of about 64 pounds of plants.

$$\begin{array}{r} 32 \\ +32 \\ \hline 64 \end{array}$$

Problem Solving Decision Making

Name_____

Use Addition and Subtraction to Make Decisions

Plan a bus route from your house to school.

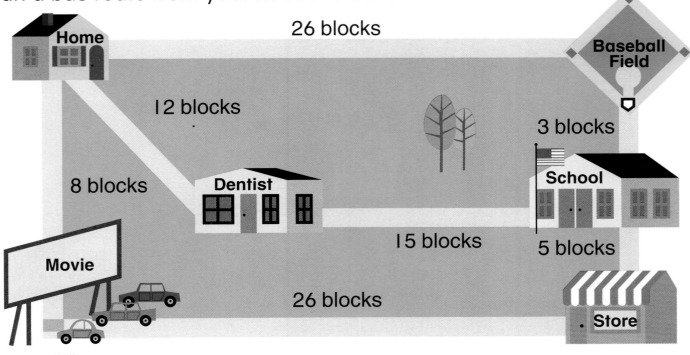

 ① Plan a route to school.
Draw your route on the map.

② How many blocks long is the route?

③ Write an addition sentence about the route.

_____

Unit 4 Decision Making

Plan 2 different routes from home to the 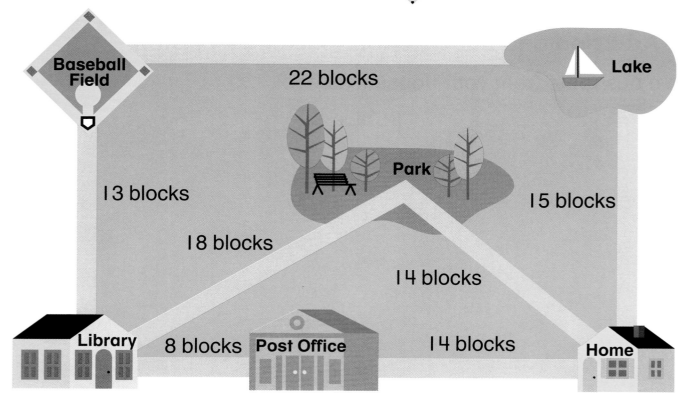.

 ④ Plan one route to the ◆.
Use a ✏ to show the route.

⑤ Plan another route to the ◆.
Use a ✏ to show the route.

 ⑥ How long is each route?

✏ _____ blocks

✏ _____ blocks

⑦ Is one route longer than the other? Write a subtraction sentence about the routes.

_____ blocks − _____ blocks = _____ blocks

Math at Home: Your child applied addition and subtraction to make decisions.
Activity: Show your child a road atlas of your state. Point out small distances between towns on the atlas. Have your child add to find the total distance.

Name _____

Unit 4 Study Guide and Review

Math Words

Draw lines to match.

1. You have to _____ when you add 47 and 36.

2. 47 + 36 is about 90.

3. 47 + 36 = 83

estimate

number sentence

regroup

Skills and Applications

Add 2-Digit Numbers (pages 231–245, 253, 262)

Examples

Count on to add. 43 + 30 = 73

Start at 43. Count on by tens.

Say 53, 63, 73.

4. 66 + 30 = _____

5. 20 + 27 = _____

6. 39 + 30 = _____

Add. Regroup if you need to.
First add the ones.
Then add the tens.

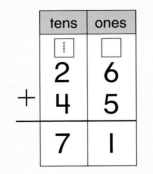

7. 59
  + 7

8. 18
  +37

9. 38
  + 6

10. 43
   +19

Unit 4 Study Guide and Review

three hundred eleven **311**

Skills and Applications

Subtract 2-Digit Numbers (pages 271–284, 293–302)

Examples

Count back to subtract.

77 − 30 = 47

Start at 77.

Count back by tens.

Say 67, 57, 47.

11 43 − 20 = _____

12 68 − 10 = _____

13 82 − 30 = _____

Subtract. Regroup if you need to.

Subtract the ones.

Then subtract the tens.

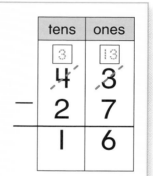

Decide if there are enough ones to subtract.

14 56
      −29

15 98
      −41

Problem Solving — Strategy

(pages 263–264, 303–304)

Circle add or subtract. Then solve.

Jack picks 14 apples. Then he picks 17 apples. How many apples does he have in all?

(add)

subtract

```
   14
 + 17
 ----
   31
```

16 Mindy has 18 leaves and 35 acorns. How many more acorns than leaves does Mindy have?

add

subtract

Math at Home: Your child practiced adding and subtracting 2-digit numbers.
Activity: Have your child use these pages to review addition and subtraction.

312 three hundred twelve

Name_____

Unit 4 Performance Assessment

Toy Sale

If you had 85¢, which two toys could you buy?

_____

_____

How much is the total? _____ ¢

How much change would you have left? _____ ¢

Use ▭▭ and ▫ or numbers to show how you solve.

Show your work.

You may want to put this page in your portfolio.

Unit 4
Enrichment

Solve a Simple Equation

There are 24 cows in all. 18 cows are outside. The rest are in the barn. How many cows are inside the barn?

First

I can write a number sentence to solve.

$18 + \boxed{} = 24$

Next

I can use subtraction to find the missing number.

$24 - 18 = 6$ So $18 + \boxed{6} = 24$.

6 cows are inside the barn.

You can use models to solve.

Solve.

Show your work.

1. There are 32 chickens in all. 15 chickens are outside. The rest are in the barn. How many chickens are inside the barn?

   _____ chickens

2. There are 44 eggs in all. 26 eggs are on the table. The rest are in the basket. How many eggs are inside the basket?

   _____ eggs

UNIT 5
CHAPTER 17

Estimate and Measure Length

ANIMAL TRACKS and FOOTPRINTS

Story by Jill Pearson
Illustrated by Gregg Valley

Next to these bird tracks I put prints of my own feet.

Are the bird's tracks longer or shorter than my footprints? _____

We find dog tracks on a hill.
We measure them with a shoe from Bill.
Who do you think has the longer foot?

Bill   dog

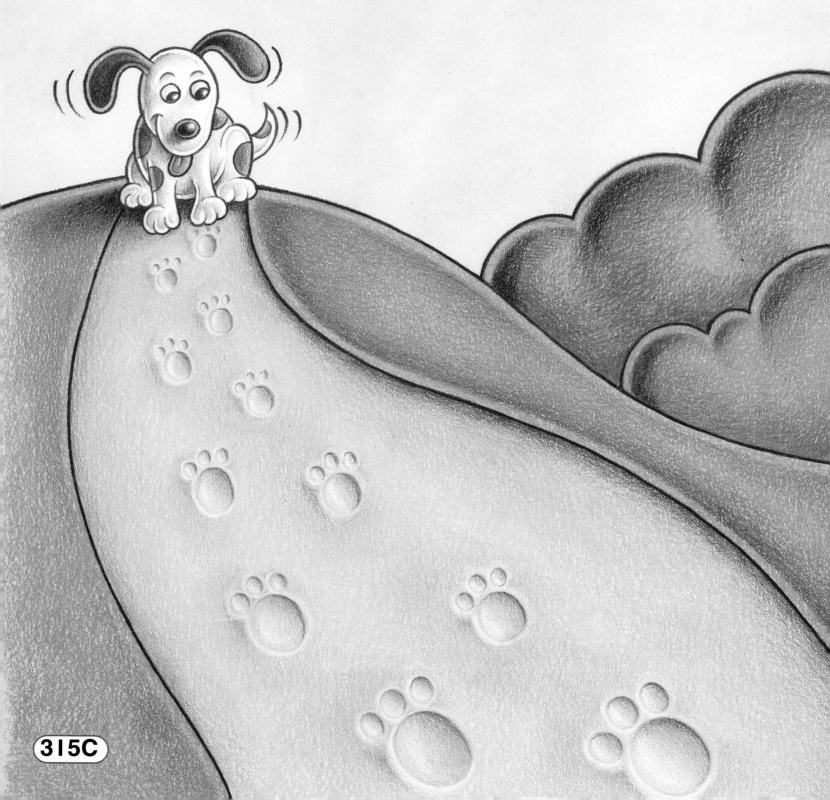

At the museum there is a footprint of a dinosaur.
It stood as tall as our classroom door.
Who do you think has the longer footprint?

you   dinosaur

Math at Home

Dear Family,

I will learn about measuring in different ways in Chapter 17. Here are my math words and an activity that we can do together.

Love, _____

My Math Words

length :
how long something is

inch :

foot :
12 inches = 1 foot

yard :
3 feet = 1 yard

distance :
how far apart two objects or places are

www.mmhmath.com
For Real World Math Activities

Home Activity

Ask your child to find household items that are longer than his or her foot. Then ask your child to find items that are shorter than his or her foot. Repeat by having your child find items longer than or shorter than his or her hand.

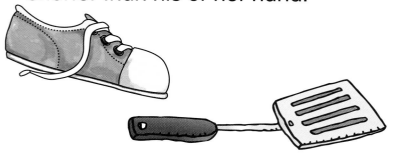

Books to Read

Look for these books at your local library and use them to help your child learn about length.

- **Carrie Measures Up** by Linda Williams Aber, The Kane Press, 2001.
- **Length** by Henry Pluckrose, Children's Press, 1995.
- **How Big Is A Foot?** by Rolf Myller, Dell Publishing Company, 1991.

Name_____

Nonstandard Units of Length

Learn Use ● to measure how long the crayon is.

"I can estimate first. I think the crayon is about 3 ● long."

Math Words
measure
estimate

Use ● to measure.
Start at the left end.
Put counters side by side. Count how many.

The crayon is about __3__ ● long.

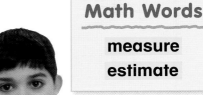

Your Turn Estimate. Then use ● to measure.

1.

   Estimate: about __7__ ●   Measure: about __7__ ●

2.

   Estimate: about _____ ●   Measure: about _____ ●

3.

   Estimate: about _____ ●   Measure: about _____ ●

4. ✏️ **Write About It!** How did you use counters to measure a pencil?

Practice Estimate. Then use small 📎 to measure.

5.

Estimate: about _____ 📎   Measure: about _____ 📎

Start at the left end to measure.

6.

Estimate: about _____ 📎   Measure: about _____ 📎

7.

Estimate: about _____ 📎   Measure: about _____ 📎

8.

Estimate: about _____ 📎   Measure: about _____ 📎

Problem Solving **Critical Thinking**

9. Measure with 🧊.
   Then measure with 📎.

   about _____ 🧊   about _____ 📎

   Are your answers the same or different?

   Explain why. _____

Math at Home: Your child used common objects as units to measure length.
Activity: Have your child estimate and then measure the length of a book in pennies.

Name _____ **Inch, Foot, and Yard**

Learn
You can use a yardstick to measure the length of longer objects.

"The toy dinosaur is about 2 feet long."

Math Words
foot (ft)
yard (yd)

1 foot = 12 inches
1 yard = 3 feet

Your Turn
Find objects like the ones shown.
Estimate the length or height.
Then use a yardstick to measure.

Object	Estimate	Measure
1. (desk)	about _____	about _____
2. 	about _____	about _____
3. (scissors)	about _____	about _____

4. **Write About It!** How many inches are in 3 feet? How do you know?

Chapter 17 Lesson 3   three hundred twenty-one **321**

Practice Find an object about as long as the estimate. Then measure.

12 inches = 1 foot
3 feet = 1 yard

Estimate	Name Your Object	Measure
5. about 1 inch	_____	about _____
6. about 8 inches	_____	about _____
7. about 1 foot	_____	about _____
8. about 2 feet	_____	about _____
9. about 2 feet	_____	about _____

Problem Solving **Mental Math**

10. Timmy's shoe is 9 inches long. His brother's shoe is 7 inches long. How much longer is Timmy's shoe?

_____ inches

11. A blue rug is 5 feet long. A green rug is 2 feet long. What is the difference between the length of the two rugs?

_____ feet

Math at Home: Your child learned about inches, feet, and yards.
Activity: Have your child show you some items in your home that are about 1 inch, 1 foot, and 1 yard long.

Name_____

Problem Solving Skill
Reading for Math

Dinosaur World

The children see a dinosaur mural. Three dinosaurs in the mural measure 2 feet long, 25 inches long, and 19 inches long.

1 foot = 12 inches

 Compare and Contrast

1. How are the dinosaurs alike? How are they different? Explain.

   _____

   _____

2. Write the lengths from greatest to least.

   _____

3. The mountains in the mural measure 26 inches, 22 inches, 23 inches, and 25 inches tall. Which are greater than 2 feet tall?

   _____

Measuring the Mural

The children want to measure other things in the mural. Three rocks measure 8 inches, 10 inches, and 12 inches long. Then they use a yardstick to measure the length of the whole mural.

1 foot = 12 inches
1 yard = 36 inches

Compare and Contrast

4. How are the rocks in the mural alike? How are they different? Explain.

_____

_____

_____

5. Which rocks are less than a foot long?

_____

6. The mural measures 46 inches long. How much longer is it than a yard?

_____

Math at Home: Your child compared and contrasted information to answer questions.
Activity: Have your child measure the lengths of two objects, then compare and contrast the results.

Chapter 17
Review/Test

Name _____

Estimate the length.
Then use an inch ruler to measure.

Estimate: about _____ inches   Measure: about _____ inches

Estimate: about _____ inches   Measure: about _____ inches

Use a centimeter ruler to measure the length.

Measure: about _____ centimeters

Measure: about _____ centimeters

Solve.

5. Jay made a paper chain 1 yard long.
Steve made a chain 2 feet long.
Who made the longer chain?
Explain how you know.

_____

three hundred twenty-nine **329**

Spiral Review and Test Prep
Chapters 1–17

Choose the best answer.

1. Measure the length of the eraser to the nearest centimeter.

   1 centimeter ○   2 centimeters ○   3 centimeters ○   4 centimeters ○

2. 27 + 18 = ☐

   35 ○   38 ○   45 ○   48 ○

3. Write the time.

   ___:___

4. In which place is the digit 5?

   54   _____

5. Look at this number pattern.
   What number would most likely come next?
   Explain the pattern.

   3, 6, 9, 12, 15, 18

LOG ON   www.mmhmath.com
For more Review and Test Prep

UNIT 5 CHAPTER 18

Estimate and Measure Capacity and Weight

Measuring Song

SING TOGETHER

Sung to the tune of "Twinkle, Twinkle, Little Star"

How long is your desk in class?

How much water's in this glass?

How much does an apple weigh?

How cold is it out today?

When you measure, you must choose

What tool is the best to use.

Math at Home

Dear Family,

I will learn about measuring weight, capacity, and temperature in Chapter 18. Here are my math words and an activity that we can do together.

Love, _____

My Math Words

capacity:
the amount a container holds when filled

The capacity is 1 cup.

temperature:
You can use temperature to measure hot or cold.

The temperature is 80°F.

www.mmhmath.com
For Real World Math Activities

Home Activity

Show your child three or four household objects. Ask your child to arrange the objects in order from lightest to heaviest.

Books to Read

Look for these books at your local library and use them to help your child learn about capacity, weight, and temperature.

- **Counting on Frank** by Rod Clement, Gareth Stevens Publishing, 1991.
- **Lulu's Lemonade** by Barbara deRubertis, The Kane Press, 2000.
- **Room For Ripley** by Stuart J. Murphy, HarperCollins, 1999.

Name_____

Explore Capacity

Learn You can use a paper cup to measure **capacity**.

I fill the large cup with dried rice.

Math Word
capacity

You can fill containers to find the amount they hold.

This small paper cup holds less rice.

Your Turn Use a large paper cup to compare. Circle more or less.

1.    more   (less)

2.    more   less

3.    more   less

4.    more   less

5.    more   less

6.    more   less

7. **Write About It!** Choose two other containers. Tell if these hold more or less rice than the large paper cup.

Chapter 18 Lesson 1   three hundred thirty-three **333**

Practice About how many paper cups does each container hold? You can use pasta or rice to measure.

I can measure to find how many paper cups will fill the bowl.

Container	Estimate	Measure
8		
9		
10		

Problem Solving — **Estimation**

11. You can use a chalkboard eraser to measure weight. Hold the eraser in one hand. Hold each object in the other hand. Tell if the object weighs more or less than the eraser.

more   less   more   less

Math at Home: Your child learned about capacity.
Activity: Choose three different-sized containers. Have your child use a cup to fill each one with water. Ask which holds the most.

334 three hundred thirty-four

Name _____

Milliliter and Liter

Learn A liter and a milliliter are units of capacity.

This bottle holds 1 liter.

A liter is a little more than 1 quart. There are 1,000 milliliters in 1 liter.

This medicine dropper holds 1 milliliter.

Math Words
liter (L)
milliliter (mL)
capacity

Your Turn Circle the better estimate. You can measure to check.

more than 1 liter

(less than 1 liter)

more than 1 milliliter

less than 1 milliliter

more than 1 liter

less than 1 liter

more than 1 milliliter

less than 1 milliliter

  Write About It! Does a drinking glass hold more or less than 1 liter? Explain.

Practice Circle the better estimate. You can measure to check.

There are 1,000 milliliters in 1 liter.

6.

   (about 2 liters)

   about 20 liters

7.

   about 5 milliliters

   about 50 milliliters

8.

   about 2 liters

   about 20 liters

9.

   about 2 liters

   about 20 liters

 Algebra • Missing Addends

Draw a picture to solve. Write a number sentence.

Show Your Work

10. Donna filled 5 glasses with 1 liter of juice. How many glasses could she fill with 3 liters of juice?

    ____ + ____ + ____ = ____

 Math at Home: Your child learned about milliliters and liters.
Activity: Have your child find one container that holds more than 1 liter and one container that holds less than 1 liter.

Name_____ **Gram and Kilogram**

Learn You can measure mass in grams and kilograms.
There are 1,000 grams in 1 kilogram.

Math Words

gram (g)
kilogram (kg)

less than
1 kilogram

about the same
as 1 kilogram

more than
1 kilogram

Your Turn Circle the better estimate.

lighter than 1 kilogram

(heavier than 1 kilogram)

lighter than 1 kilogram

heavier than 1 kilogram

lighter than 1 kilogram

heavier than 1 kilogram

lighter than 1 kilogram

heavier than 1 kilogram

  Write About It! Is a large object always heavier than a small object? Explain.

Practice Circle the better estimate. Then use a balance scale to measure.

An apple is less than 1 kilogram.

Object	Estimate	Measure
6. (dictionary)	lighter than 1 kilogram / (heavier than 1 kilogram)	lighter than 1 kilogram / (heavier than 1 kilogram)
7. (pears)	lighter than 1 kilogram / heavier than 1 kilogram	lighter than 1 kilogram / heavier than 1 kilogram
8. (banana)	lighter than 1 kilogram / heavier than 1 kilogram	lighter than 1 kilogram / heavier than 1 kilogram
9. (quarter)	lighter than 1 kilogram / heavier than 1 kilogram	lighter than 1 kilogram / heavier than 1 kilogram

Problem Solving Critical Thinking

10. Which is heavier, 1 kilogram of rocks or 1 kilogram of feathers? Explain.

_____

_____

Math at Home: Your child learned about grams and kilograms.
Activity: Have your child name one object that is less than a kilogram and one object that is more than a kilogram.

342 three hundred forty-two

Name_____

Temperature

Learn Temperature can be measured in degrees Fahrenheit (°F).

Math Words
temperature
degrees Fahrenheit (°F)
degrees Celsius (°C)

This thermometer shows __60__ °F.

This thermometer shows __82__ °F.

Try It Write each temperature.

1.

__55__ °F

2.

_____ °F

3. **Write About It!** What do you think the temperature is today? Look at the thermometer to check.

Chapter 18 Lesson 6 three hundred forty-three **343**

Practice Write each temperature.

Temperature can be measured in degrees Celsius (°C), too.

   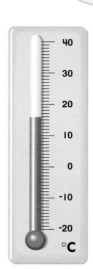 

4) _18_ °C 5) _____ °C 6) _____ °C 7) _____ °C

Problem Solving — Critical Thinking

Show Your Work

8) Read the thermometer. Draw a picture. Show what you would wear to go outside. Write about your picture.

_____

_____

_____

_____

_____

Math at Home: Your child learned about temperature.
Activity: Ask your child what you should wear when the temperature is 25 °F.

344 three hundred forty-four

Name _____

Use Logical Reasoning

You can use logical reasoning to help you solve problems. Carrie wants to know the temperature outside. She needs to decide which tool to use.

Problem Solving Strategy

cup   balance scale   thermometer

Read

What do I already know? _____

What do I want to find out? _____

What do I need to find out? _____

Plan

I can decide which tool measures temperature.

Solve

I can carry out my plan.

A _____ measures temperature.

Look Back

Does my answer make sense?   Yes   No

Why? _____

Chapter 18 Lesson 7   three hundred forty-five **345**

Circle the correct tool to measure.

1 How heavy is it?

2 How cold is it?

3 How much water does it hold?

4 How heavy is it?

Math at Home: Your child learned about measurement tools.
Activity: Ask your child what tools he or she would use to measure different items in your home.

346 three hundred forty-six

Game Zone

Practice at School ★ Practice at Home

Name_____

How Big?

▶ Each partner chooses a ruler.
▶ Find one object to match each length.
▶ Write what you found.
▶ The first player to find 3 matching lengths wins.

 2 players

You Will Need

inch ruler

centimeter ruler

Length	Found Object about this Size
Player 1	
about 2 inches	
about 7 inches	
about 4 inches	
Player 2	
about 2 cm	
about 7 cm	
about 18 cm	

Chapter 18 Game Zone

three hundred forty-seven **347**

Technology Link

Addition and Subtraction • Calculator

Use a to find the new temperature.

You Will Use

The temperature is 62°F.
It goes down 9°F.
What is the new temperature?

Press.

You see .

The new temperature is 53°F.

---

1 The temperature is 95°F.
It goes down 6°F.
What is the new temperature?

Press. _____ _____ _____

_____°F is the new temperature.

---

2 The temperature is 58°F.
It goes up 6°F.
What is the new temperature?

Press. _____ _____ _____

_____°F is the new temperature.

3 ✏️ **Write About It!** Why would you use a calculator or mental math to solve these problems?

Chapter 18
Review/Test

Name_____

Circle the better estimate.

more than 1 ounce
less than 1 ounce

more than 1 cup
less than 1 cup

more than 1 pound
less than 1 pound

more than 1 kilogram
less than 1 kilogram

about 2 liters
about 200 liters

about 2 milliliters
about 200 milliliters

Write each temperature.

_____ °F

_____ °C

9. Circle the tool you would use to measure how heavy a shoe is.

thermometer   cup   scale   centimeter ruler

10. Danny fills a jug with 4 pints of water. Does he have more or less than 1 quart of water? Explain your answer.

_____

Chapter 18 Review/Test

three hundred forty-nine **349**

Spiral Review and Test Prep
Chapters 1–18

Choose the best answer.

1) There are 72 oranges in the box. The clerk takes out 27 oranges to make juice. How many oranges are left?

- 45 oranges
- 55 oranges
- 85 oranges
- 99 oranges

2) Jan has 49¢. Which group of coins could she have?

- 8 nickels, 4 pennies
- 1 quarter, 2 dimes, 4 pennies
- 1 quarter, 2 dimes, 9 pennies
- 2 quarters

3) A cat weighs about ____.

- 5 ounces
- 10 ounces
- 9 pounds
- 100 pounds

4) Name an item that is about 1 meter long. _____

5) Write the number that makes the number sentence true.   56 = ____ + 10

6) Sue started her homework at 3:30. She finished at 5:30. How long did she spend doing homework? Explain your answer.

_____

_____

LOG ON www.mmhmath.com
For more Review and Test Prep

350 three hundred fifty   Spiral Review and Test Prep

UNIT 5 CHAPTER 19
2-Dimensional and 3-Dimensional Shapes

A Riddle

I'm shaped just like a grapefruit half,
But upside-down and frozen.
I'm big and hollow, with a door
In case somebody goes in.

I'm dry inside when icy storms
Outside make people shiver.
Enter, and you'll soon enjoy
The heat that I deliver.

I'm seen in snowy places.
I warm up and keep dry
Whoever walks inside me.
Can you tell me: Who am I?

Answer: igloo

Math at Home

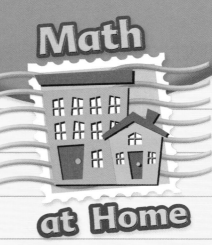

Dear Family,

I will learn about figures and ways to describe them in Chapter 19. Here are my math words and an activity that we can do together.

Love, _____

My Math Words

3-Dimensional Figures:

rectangular prism

sphere   cylinder   pyramid

2-Dimensional Shapes:

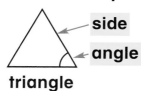
triangle

parallelogram   hexagon

www.mmhmath.com
For Real World Math Activities

Home Activity

Place some objects that are solid shapes on a table.

Choose an object and give several clues to its identity. For example: I am round. I am flat. I have no corners. What am I?

Take turns describing and identifying different objects.

Books to Read

Look for these books at your local library and use them to help your child learn about shapes and figures.

- **Kitten Castle** by Mel Friedman and Ellen Weiss, The Kane Press, 2001.
- **Round Is a Mooncake** by Roseanne Thong, Chronicle Books LLC, 2000.
- **Round and Square** by Miriam Schlein, Mondo Publishing, 1999.

Name_____

2-Dimensional and 3-Dimensional Relationships

HANDS ON Activity

Learn Find a 2-dimensional shape in a 3-dimensional figure. You can trace a face.

All the faces of the cube are squares.

Your Turn Find objects like these 3-dimensional figures. Trace around the bottom face. Circle the 2-dimensional shape you made.

 1
 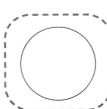

2
 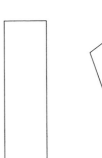

 3

4

 5 **Write About It!** You traced around the face of a 3-dimensional figure to make a circle. What 3-dimensional figure did you trace?

Chapter 19 Lesson 3

Practice Circle the 3-dimensional figures you can use to make all the 2-dimensional shapes.

2-Dimensional Figures	3-Dimensional Figures
6. rectangle, circle, circle	(cylinder) ⭕, pyramid
7. six squares	cube, rectangular prism
8. four triangles, square	cube, pyramid
9. rectangles	rectangular prism, rectangular prism

Problem Solving — Visual Thinking

10. Compare the ball and the circle. How are they alike and different?
_____
_____

11. Compare the block and the square. How are they alike and different?
_____
_____

Math at Home: Your child learned about the shapes of faces of 3-dimensional objects.
Activity: Have your child trace the faces of a box and a can and name the 2-dimensional shapes he or she traced.

Name_____

Combine Shapes

HANDS ON Activity

 You can put two shapes together to make a new shape.

A trapezoid has 4 sides. A hexagon has 6 sides.

Math Words
trapezoid
hexagon

I can use 2 trapezoids to make a hexagon.

trapezoid

hexagon

Your Turn Use pattern blocks to make new shapes. Then complete the chart.

Use these pattern blocks.	Draw a new shape.	How many sides?	How many angles?	Name of new shape.
1. trapezoid + trapezoid	hexagon	6	6	hexagon
2. triangle + triangle		___	___	

3.   **Write About It!** Can you use squares to make a rectangle? Explain.

Chapter 19 Lesson 4

three hundred fifty-nine **359**

Practice Use pattern blocks to make new shapes. Complete the chart.

Use these pattern blocks.	Make a new shape.	How many sides?	How many angles?	Name of new shape.
4 (two squares)	(rectangle outline)	4	4	rectangle
5 (triangle and rhombus)				
6 (trapezoid and two triangles)				

Problem Solving — Visual Thinking

Show Your Work

7. Tommy made this shape using 3 pattern blocks. What were the blocks he used? Draw lines to show how Tommy put the blocks together.

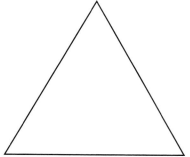

Math at Home: Your child combined shapes to make new shapes.
Activity: Have your child show you how to use 2 triangles to make a new shape.

360 three hundred sixty

Extra Practice

Name _____

Add or subtract. Circle the ones that need regrouping.

46
− 23

23
+17
40

60
− 12

74
− 18

49
+11

26
− 19

36
+36

18
+21

16
+34

40
− 15

Draw the hour and minute hands to show the time.

1.
2.
3.
4.
5.
6.
7.
8.
9.
10.

www.mmhmath.com
LOG ON For more Practice

 Math at Home: Your child practiced telling time.
Activity: Ask your child to tell you when it is 2:50. Repeat with different times.

Name_____ **Shape Patterns**

Learn You can make a pattern of shapes. Use pattern blocks. Then use letters to show the pattern another way.

Math Word
unit

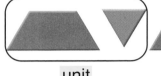
unit

(A B) A B A B A B A B

To find a pattern, look for the shapes that repeat. The shapes that repeat make a unit.

Your Turn Use pattern blocks to show the pattern. Then circle the pattern unit.

1.

2.

3.

4.

5.

6. **Write About It!** How can you tell which shapes come next?

Practice Use pattern blocks to show the pattern. Then use letters to show each pattern another way.

7) ▲ ▼ ▲ ▼ ▲ ▼ ▲ ▼
A B A B A B A B

8)
___ ___ ___ ___ ___ ___ ___ ___ ___

9) ■ ▼ ◢ ■ ▼ ◢ ■ ▼ ◢
___ ___ ___ ___ ___ ___ ___ ___ ___

10) ⬡ ▲ ⬡ ⬡ ▲ ⬡ ⬡ ▲ ⬡
___ ___ ___ ___ ___ ___ ___ ___ ___

 Make it Right

Draw a picture to solve.

11) What is wrong with Meg's pattern? Circle the shape that is wrong.

Draw the correct pattern below.

 Math at Home: Your child learned about shape patterns.
Activity: Have your child draw a shape pattern, using squares and triangles.

Name _____

Problem Solving Skill
Reading for Math

The City

Min and Anna build a model of their city. They use 3-dimensional figures for the buildings and trees. They use 2-dimensional shapes for signs.

 Use Illustrations

Use the picture to help answer the questions.

1. Name two 3-dimensional figures that were used in the city. How are they alike? How are they different?

   _____

   _____

   _____

2. Which 2-dimensional shapes were used for the signs in the city?

   _____

3. How are the signs the same? How are they different?

   _____

   _____

Chapter 19 Lesson 6

three hundred sixty-five **365**

The Museum and the Park

Anna made a museum and park in the city. How many different kinds of shapes did she use?

Problem Solving

 Use Illustrations

Use the picture to help answer the questions.

4 Name the 2-dimensional shapes that were used in the park.

_____

_____

5 Find the 2-dimensional sign. What shape is it?

_____

6 Choose 2 of the 2-dimensional shapes you named. Tell how they are the same and how they are different.

_____

_____

_____

 Math at Home: Your child used illustrations to help him or her answer questions.
Activity: Ask your child to draw some shapes, then use the pictures to explain how the shapes are alike and different.

Name _____

Problem Solving Practice

Solve.

1 Kathy painted this picture.

How many ▢ ? _____

How many △ ? _____

2 Chris and Beth painted this picture of a bus. Look for the 2-dimensional shapes in the picture.

How many squares? _____

How many circles? _____

Write a Story!

3 Write a story about the shapes in this picture.

_____

_____

_____

_____

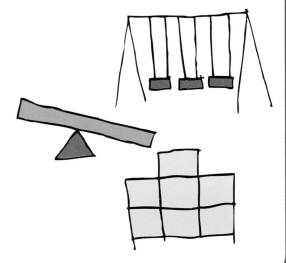

Chapter 19 Problem Solving Practice

three hundred sixty-seven **367**

Writing for Math

 Write about the object in the picture.
Use the words on the cards to help.

| faces | vertices | edges |

Think

I can tell the shape of the object.
What do I know about 3-dimensional figures?

_____

_____

Solve

I can look at the object and name the figure.

_____.

Explain

I can tell you how I know the figure.

_____

_____

_____

Chapter 19
Review/Test

Name _____

1 Circle the 3-dimensional figure.

2 Circle the 2-dimensional shape.

3 Name the 3-dimensional figure.
Write how many faces, vertices, and edges.

   Name: _____   Faces: _____

Vertices: _____   Edges: _____

4 Which 2-dimensional shape could you make
if you traced the 3-dimensional figure?
Circle it. Name the 2-dimensional shape.

     _____

5 Use pattern blocks to show the pattern.
Then use letters to show the pattern
another way.

___ ___ ___ ___ ___ ___ ___ ___ ___

Spiral Review and Test Prep
Chapters 1–19

Choose the best answer.

1) Sonia had 16 crayons.
She gave 2 crayons to her brother.
How many crayons did she have left?
Which number sentence would you use?

16 − 2 = 14	6 + 2 = 8	20 + 2 = 22	2 + 16 = 18
○	○	○	○

2) Which figure is a cone?

○   ○   ○   ○

3) Which unit would you use to measure how much water a bathtub holds?

foot	liter	pound	milliliter
○	○	○	○

4) I am a number between 70 and 90.
When you count by 10, you say my name.
What number am I? _____

5) Todd had 17 crackers. He ate 3 of them.
How many crackers were left?
Bobby says the answer to this problem is
20 crackers. Is he correct? Tell how you know.

370 three hundred seventy

UNIT 5

CHAPTER 20

Spatial Sense

READ TOGETHER

Paper Dolls

It's no surprise the shape and size
Of these dolls is the same.
My problem now, to figure how
To give each one a name!

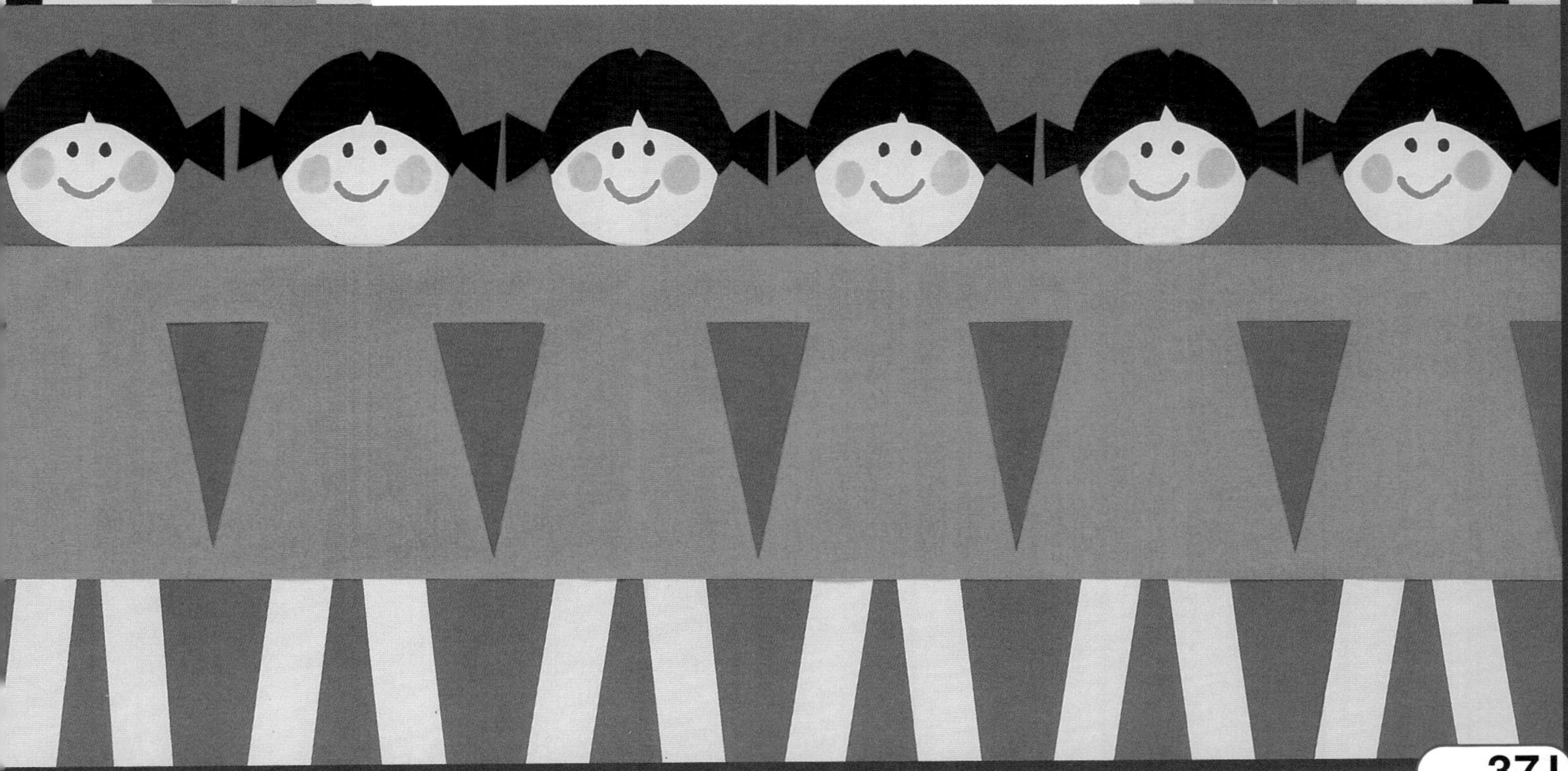

Math at Home

Dear Family,

I will learn more about figures and ways they are related in Chapter 20. Here are my math words and an activity that we can do together.

Love, _____

My Math Words

congruent :
same size and shape

line of symmetry :
line along which a shape can be folded so that the two parts match exactly

perimeter :

1 + 2 + 1 + 2 = 6 units

area :
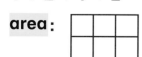
The area is 6 square units.

www.mmhmath.com
For Real World Math Activities

Home Activity

Place some 3-dimensional objects on a table.

Have your child make shape pictures by tracing around the faces of the objects. Take turns telling each other about the shapes in the pictures.

Books to Read

In addition to these library books, look for the Time For Kids math story that your child will bring home at the end of this unit.

- **Racing Around** by Stuart J. Murphy, HarperCollins, 2002.
- **Spaghetti and Meatballs for All!** by Marilyn Burns, Scholastic Press, 1997.
- **Time for Kids**

Name_____

Slides, Flips, and Turns

Learn You can move shapes in different ways.

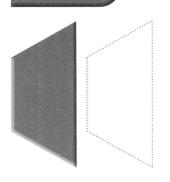  

This is a **slide**.   This is a **flip**.   This is a **turn**.

Math Words
slide
flip
turn

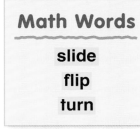
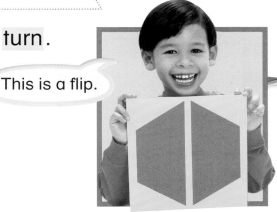

This is a flip.

Your Turn Color to show the move. Use the pattern blocks.

1. Which shows a slide?

 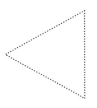 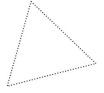 

2. Which shows a flip?

  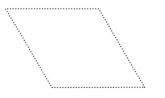 

3. Which shows a turn?

  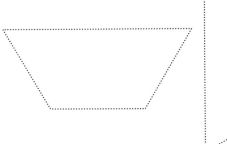

4.   Write **About It!** How are a flip, slide, and turn alike?

Chapter 20   Lesson 3   three hundred seventy-seven **377**

Practice Write the word that names the move.

slide	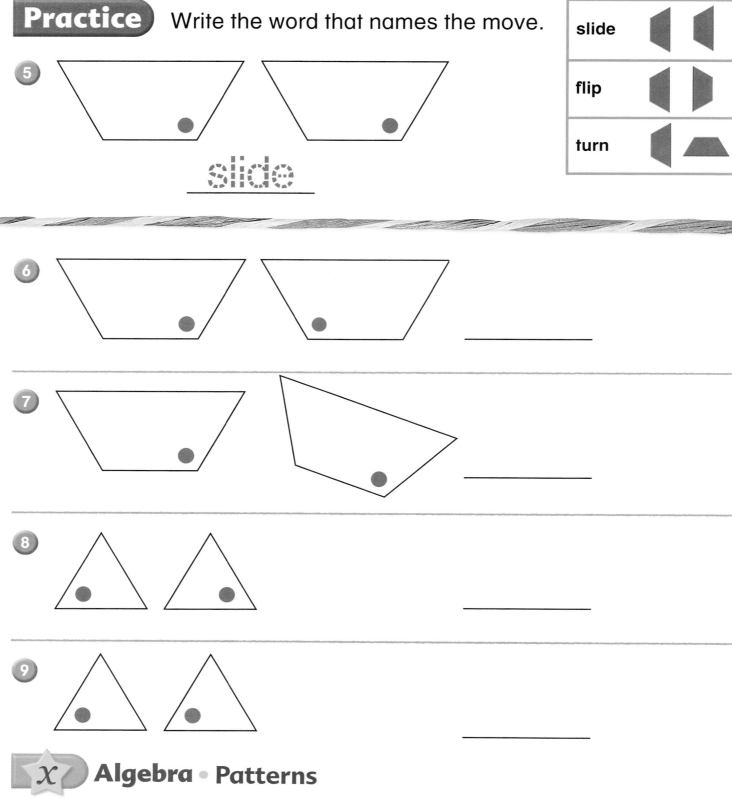
flip	
turn	

5.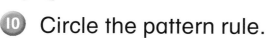
   slide

6. _____

7. _____

8. _____

9. _____

Algebra • Patterns

10. Circle the pattern rule.

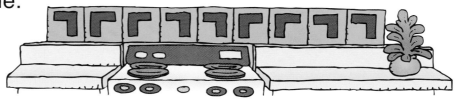

flip, slide   slide, turn   turn, slide

 Math at Home: Your child learned how to slide, flip, and turn shapes.
Activity: Cut out a paper triangle and have your child slide, flip, and turn this shape.

378 three hundred seventy-eight

Name_____

Game Zone
Practice at School ★ Practice at Home

Build It

▸ Take turns. Each player covers part of the picture with one of the pattern blocks.

▸ Continue until the picture is completed.

▸ The winner is the player who places the last pattern block on the picture.

 2 players

You Will Need
pattern blocks

Chapter 20 Game Zone three hundred eighty-five **385**

Technology Link

Shapes • **Computer**

Use 🔷 to make shapes.

- Choose a mat to show shapes.
- Stamp out 2 ▲.
- Turn 1 ▲ 2 times.
- Put the ▲▲ together.

  You made a parallelogram.

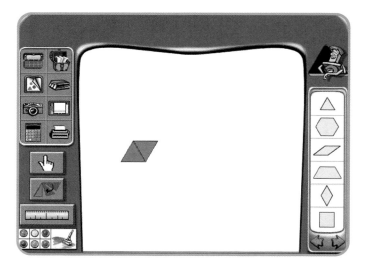

Make other shapes and put them together.
You can use the computer or pattern blocks.

1. Stamp out 2 ▰.

   Flip 1 ▰.

   Slide 1 ▰.

   Show what you made.

2. Stamp out 3 ▰.

   Turn 1 ▰.

   Flip 1 ▰.

   Show what you made.

 For more practice use Math Traveler.™

Chapter 20
Review/Test

Name _____

① Find the area in square units.

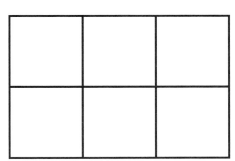 _____ square units

② Color the two congruent shapes.

③ Draw a line of symmetry.

④ Circle the move.

 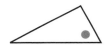

slide   flip   turn

Guess and check to solve.

⑤ Jill draws a square. The perimeter of the square is 12 centimeters. How long is each side of the square?

_____ centimeters

Spiral Review and Test Prep
Chapters 1–20

Choose the best answer.

1 Which number makes this sentence true?

$6 + \square = 14$

- 7
- 8
- 10
- 20

2 Which are even numbers?

- 16, 18, 20
- 16, 17, 20
- 17, 19, 21
- 18, 20, 21

3 Write the number that comes just between 63 and 65.

_____

4 Square units are used to measure _____ .

5 Henry has a quarter and 3 pennies. He finds 6 more pennies. How much money does Henry have now? You can draw a picture or use numbers to show how you solve.

_____ cents

TIME FOR KIDS

Name _____

At this museum, you can see two giant spheres. What shapes do you see in the picture? Draw them in the space below.

Museum Math

TIME FOR KIDS

Museums are fun places to visit. You can learn many things from them.

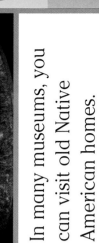

In many museums, you can visit old Native American homes. What shapes do you see?

This outdoor museum has butterflies! The butterflies show symmetry.

One museum has the machine that took astronauts to the moon. They measured what they found there.

Name_____

Linking Math and Science

Weight and Soils

The Earth is covered with different kinds of soil.

Science Words
topsoil
sandy soil

Topsoil

Topsoil is dark brown or black. It can hold water.

Sandy Soil

Sandy soil is light in color. It does not hold water well.

Problem Solving

Use the pictures.
Circle the word to complete each sentence.

1. A soil that doesn't hold water well is ____.

    topsoil   sandy soil

2. A soil that can hold water is ____.

    topsoil   sandy soil

Unit 5 Linking Math and Science

What to Do

- **Observe** Tell how each soil feels.

  _____

  _____

  _____

- Fill a cup with sandy soil. Label it.
- Fill the other cup with topsoil. Label it.
- **Measure** Put each cup on the balance.

Solve.

3 **Compare** Which soil is heavier?

4 **Infer** Why do you think that soil is heavier?

_____

_____

_____

Math at Home: Your child applied measurement concepts to investigate properties of soil.
Activity: Repeat this activity with your child using different kinds of soils from around your home.

Name_____

Unit 5
Study Guide and Review

Math Words

Draw lines to match.

1. 4 cups --------- 1 foot
2. 48° F ---------- 1 quart
3. 12 inches ------ temperature

Skills and Applications

Measurement (pages 317–322)

Examples

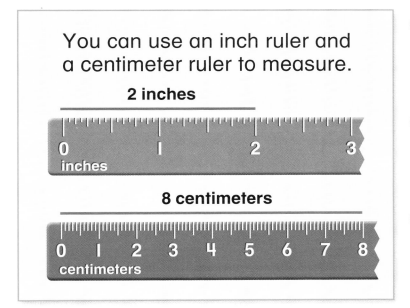

You can use an inch ruler and a centimeter ruler to measure.

4. _____ centimeters

5. _____ centimeters

6. _____ inches

You can decide the better unit of measure.

(ounce) / pound   gram / (kilogram)

7.  ounce / pound

8.  gram / kilogram

9.  ounce / pound

Unit 5 Study Guide and Review   three hundred ninety-one **391**

Skills and Applications

Geometry (pages 353–358; 377–378)

Examples

Circle the 3-dimensional figure you can use to make the 2-dimensional shape.

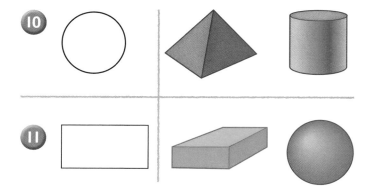

Write the word turn, flip, or slide to tell how the shape was moved.

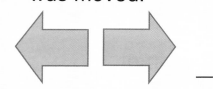

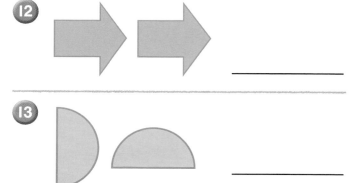

Problem Solving — Strategy

(pages 353–354)

14. Tell what you know about the 3-dimensional figure shown. Write how many faces, edges, and vertices it has.

_____   _____

Math at Home: Your child learned about measurement and geometry.
Activity: Have your child use these pages to review measurement and shapes.

Name _____

Unit 5 Performance Assessment

Drawing Fun!

Draw and measure to answer the questions.

You Will Need

or

① Draw a path 10 units long. Then draw a different path 10 units long.

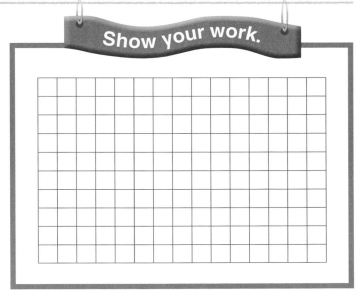

② Draw a symmetrical shape that has 4 sides. Then make a line of symmetry. Choose a ruler to measure the sides. Then add to find the perimeter.

The perimeter is _____.

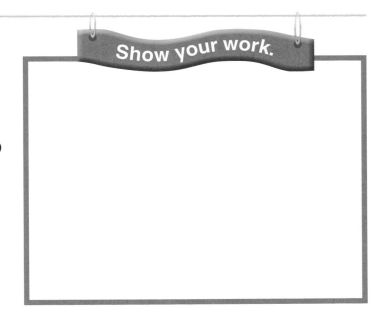

You may want to put this page in your portfolio.

Unit 5
Enrichment

What is the Volume?

Volume is the amount of space an object takes up. You can find the volume of a figure.

cubes

First Estimate how many cubes are in this figure.

Estimate: about __20__ cubes

Next Build the figure. Write how many cubes you used.

Count: There are __18__ cubes.

Estimate the number of cubes. Then build the figure. Write how many cubes you used.

1

Estimate: about ____ cubes

Count: There are ____ cubes.

2

Estimate: about ____ cubes

Count: There are ____ cubes.

3

Estimate: about ____ cubes

Count: There are ____ cubes.

4

Estimate: about ____ cubes

Count: There are ____ cubes.

UNIT 6
CHAPTER 21

Place Value to Thousands

READ TOGETHER

We Can Make Almost Anything

Story by Marina Ramos
Illustrated by Kristina Stephenson

We string the beads in a row.
Count the beads as you go.

Now there are _____ beads.

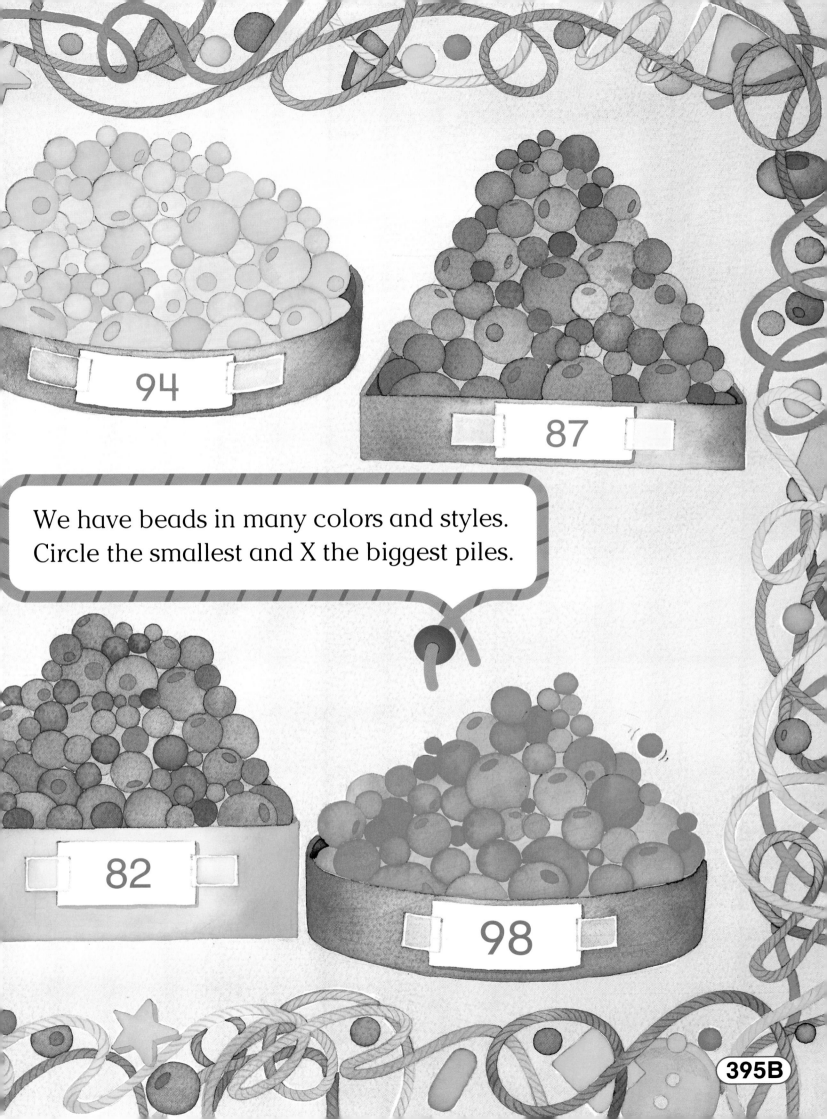

If you make these 2 strings of beads, how many beads will you need? _____ beads

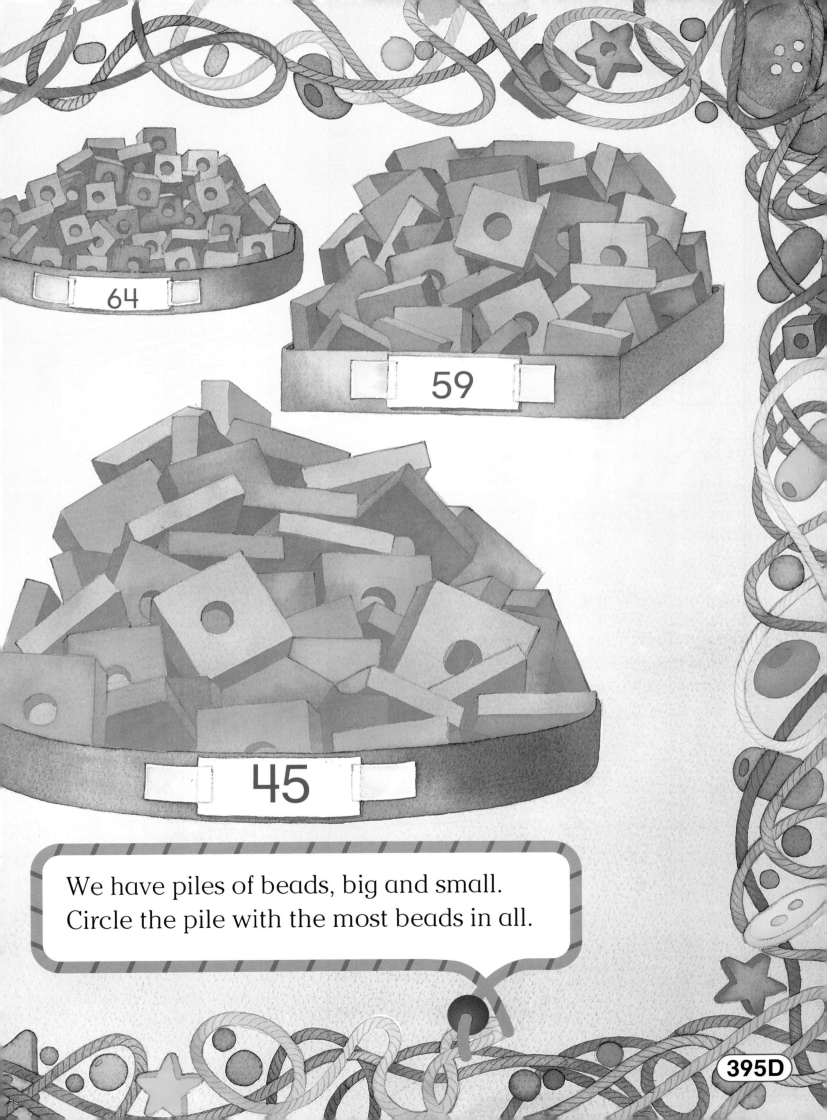

64

59

45

We have piles of beads, big and small.
Circle the pile with the most beads in all.

Math at Home

Dear Family,

I will learn about place value to thousands in Chapter 21. Here are my math words and an activity that we can do together.

Love, _____

My Math Words

digits:
used when representing a number

the digits are 1, 2, 4.

place value:
the amount that each digit in a number stands for

1 → hundreds
2 → tens
4 → ones

Home Activity

Put about 100 beans, beads, or buttons in a bowl.

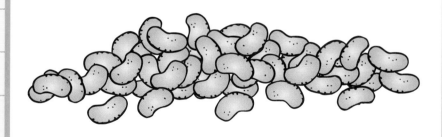

Have your child find different ways to group and count the beans, such as by 5s or 10s.

www.mmhmath.com
For Real World Math Activities

Books to Read

Look for these books at your local library and use them to help your child learn place value to thousands.

- **Only One** by Marc Harshman, Cobblehill Books, 1993.
- **The 329th Friend** by Marjorie Weinman Sharmat, Four Winds Press, 1992.
- **Moira's Birthday** by Robert N. Munsch, Firefly Books Ltd., 1988.

Name_____ **Hundreds**

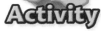

Learn You can group tens to make hundreds.

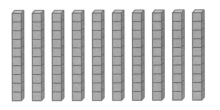 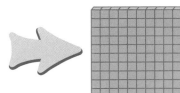

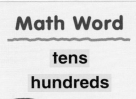

Math Word
tens
hundreds

__10__ tens = __1__ hundred = __100__ in all

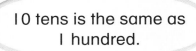

10 tens is the same as 1 hundred.

Your Turn Use models to make groups of hundreds. Write how many hundreds.

1. 20 tens = __2__ hundreds
2. 30 tens = _____ hundreds

3. 40 tens = _____ hundreds
4. 50 tens = _____ hundreds

5. 60 tens = _____ hundreds
6. 70 tens = _____ hundreds

7. 80 tens = _____ hundreds
8. 90 tens = _____ hundreds

9. Write **About It!** How many hundreds are there in 900? How do you know?

Chapter 21 Lesson 1   three hundred ninety-seven **397**

Practice Write how many tens and how many ones. Then write the number.

You can trade 1 hundred for 10 tens.

10.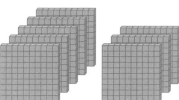
8 hundreds = 80 tens = 800 ones = 800

11.
4 hundreds = ____ tens = ____ ones = ____

12.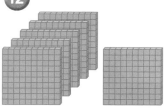
6 hundreds = ____ tens = ____ ones = ____

13.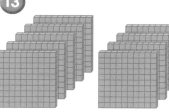
9 hundreds = ____ tens = ____ ones = ____

Problem Solving Number Sense

14. Randy trades 10 hundreds for 1 thousand.

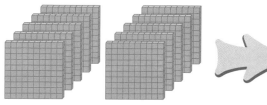

10 hundreds is 1,000.

Write the number. _____

Math at Home: Your child used models to make hundreds.
Activity: Ask your child to count by hundreds to 1,000.

Name_____

Hundreds, Tens, and Ones

 You can use hundreds, tens, and ones to show 325.

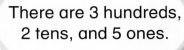 There are 3 hundreds, 2 tens, and 5 ones.

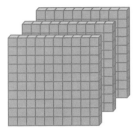 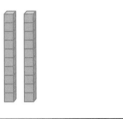

hundreds	tens	ones
3	2	5

__3__ hundreds __2__ tens __5__ ones

 Write how many hundreds, tens, and ones.

1) 458

__4__ hundreds __5__ tens __8__ ones

hundreds	tens	ones
4	5	8

2) 142

____ hundred ____ tens ____ ones

hundreds	tens	ones

3) 636

____ hundreds ____ tens ____ ones

hundreds	tens	ones

4) **Write About It!** How many hundreds, tens, and ones are in 572? Explain how you know.

Chapter 21   Lesson 2

three hundred ninety-nine **399**

Practice Write how many hundreds, tens, and ones.

Remember to write a zero when there are no tens or ones.

5) 630

__6__ hundreds __3__ tens __0__ ones

hundreds	tens	ones
6	3	0

6) 246

____ hundreds ____ tens ____ ones

hundreds	tens	ones

7) 515

____ hundreds ____ ten ____ ones

hundreds	tens	ones

Write each number.

8) 3 hundreds 2 tens 6 ones

_____

9) 5 hundreds 0 tens 8 ones

_____

Problem Solving — Number Sense

10) Look at these two numbers.

What does the 0 stand for in each number?

305   350

_____

Math at Home: Your child learned about hundreds, tens, and ones in 3-digit numbers.
Activity: Write a 3-digit number such as 291. Ask your child to tell how many hundreds, tens, and ones.

Name _____

Place Value through Hundreds

Learn You can learn the place value of a digit by its place in a number.

Math Words
- place value
- digit
- expanded form

hundreds	tens	ones
[flats image]	[rods image]	[units image]
2	5	7

2 hundreds 5 tens 7 ones

200 + 50 + 7

257

You can write a number in expanded form.

Try It Write how many hundreds, tens, and ones. Then write the number.

1) 3 hundreds 1 ten 5 ones

hundreds	tens	ones
3	1	5

300 + 10 + 5

315

2) 3 hundreds 2 tens 2 ones

hundreds	tens	ones

____ + ____ + ____

____

3) 2 hundreds 4 tens 8 ones

hundreds	tens	ones

____ + ____ + ____

____

4) 1 hundred 0 tens 7 ones

hundreds	tens	ones

____ + ____ + ____

____

5) **Write About It!** How can you tell the value of a digit in a 3-digit number?

Chapter 21 Lesson 3   four hundred one **401**

 **Practice** Write how many hundreds, tens, and ones. Then write the number.

The value of the digit 6 in 672 is 600.

6) 6 hundreds 7 tens 2 ones

hundreds	tens	ones
6	7	2

600 + 70 + 2

672

7) 8 hundreds 3 tens 1 one

hundreds	tens	ones

___ + ___ + ___

___

8) 9 hundreds 2 tens 5 ones

hundreds	tens	ones

___ + ___ + ___

___

9) 7 hundreds 4 tens 0 ones

hundreds	tens	ones

___ + ___ + ___

___

Circle the value of the green digit.

10) 591

(500) 50 5

11) 256

600 60 6

12) 924

200 20 2

13) 485

800 80 8

14) 372

200 20 2

15) 273

200 20 2

Math at Home: Your child learned the place value of 3-digit numbers.
Activity: Say a 3-digit number and have your child tell you the value of each digit.

402 four hundred two

Name _____

Compare. Use < or >.

Color > 🖍.

Color < 🖍.

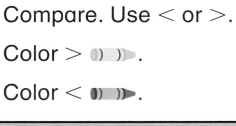

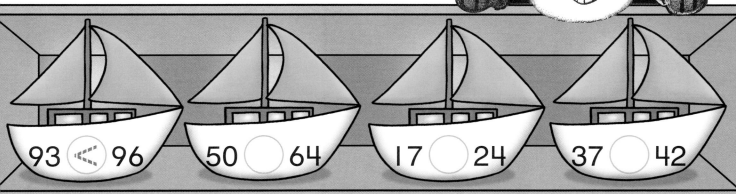

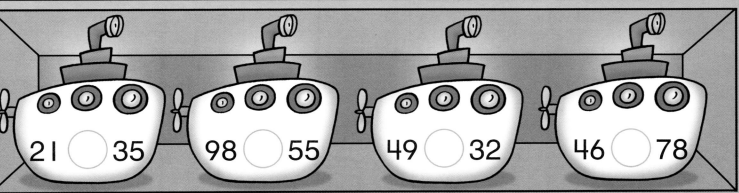

Chapter 21 Extra Practice

four hundred three **403**

Extra Practice

Write the numbers in order from least to greatest.

1) 44   22   57   40

2) 77   92   58   31

3) 33   39   11   24

4) 85   81   99   63

www.mmhmath.com
For more Practice

Math at Home: Your child practiced comparing and ordering 2-digit numbers.
Activity: Ask your child to arrange the numbers in exercise 2 on this page from greatest to least.

Name_____

Explore Place Value to Thousands

Learn You can use models and a place-value chart to explore numbers to thousands.

Math Word
thousands

1 thousand 2 hundreds 4 tens 5 ones

thousands	hundreds	tens	ones
(cube)	(flats)	(rods)	(units)
1	2	4	5

You can show numbers in different ways.

1,000 + 200 + 40 + 5   1,245

Word name: one thousand two hundred forty-five

Try It Write how many thousands, hundreds, tens, and ones. Then write the number.

1) 2 thousands 1 hundred 6 tens 8 ones

thousands	hundreds	tens	ones
2	1	6	8

2,000 + 100 + 60 + 8   2,168

---

2) 1 thousand 1 hundred 3 tens 4 ones

thousands	hundreds	tens	ones

_____ + ____ + ____ + ____   _____

3) ✏ **Write About It!** Tell what you know about the number 2,561.

Chapter 21   Lesson 4   four hundred five **405**

Practice Write how many thousands, hundreds, tens, and ones. Then write the number.

Write a zero when there are no hundreds, tens, or ones.

④ 2 thousands 1 hundred 0 tens 5 ones

thousands	hundreds	tens	ones

_____ + _____ + _____ + _____   _____

⑤ 1 thousand 3 hundreds 8 tens 2 ones

thousands	hundreds	tens	ones

_____ + _____ + _____ + _____   _____

Use the number line. Write the missing number in the pattern.

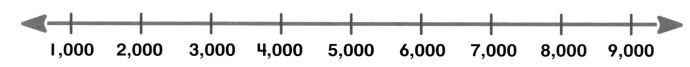

⑥ 1,000, 2,000, 3,000, _____, 5,000, 6,000, 7,000, _____, 9,000

Problem Solving **Number Sense**

⑦ Pat has 982 beads. Does she have more or less than 1,000 beads?

Explain. _____

_____

Math at Home: Your child explored numbers to the thousands.
Activity: Write a 4-digit number. Have your child tell you what he or she knows about the number.

Name_____

Problem Solving Skill
Reading for Math

Dana's Marbles

Dana collects things. She has marbles in 4 different colors. She keeps each color in a separate jar. Dana has 98 red, 136 purple, 87 yellow, and 69 pink marbles.

Reading Skill Find the Main Idea

1. What is the story about?

   _____

2. Write the numbers of marbles in order from greatest to least.

   _____

3. Write how many hundreds, tens, and ones for the number of purple marbles.

   ____ hundred ____ tens ____ ones

4. If someone gave Dana 10 more yellow marbles, how many yellow marbles would she have in all?

   _____

Chapter 21 Lesson 5

four hundred seven **407**

Coin Collectors

Dana and her friend Jack collect quarters from different states. They have 425 quarters in all. There are 25 quarters from North Carolina. The rest are from other states.

Reading Skill **Find the Main Idea**

1. What is the main idea of this story?

   _____

2. Write the word name for 425.

   _____

3. Write the number 425 in expanded form.

   _____ + _____ + _____

4. Dana has 310 quarters. Jack has 115 quarters.

   Who has more quarters? _____

Math at Home: Your child identified a main idea to answer questions.
Activity: Read a story with your child. Ask him or her to tell you a short sentence describing what the story is mainly about.

Name _____

Problem Solving Practice

Solve.

1) There are 217 crayons in the box.

Write how many hundreds, tens, and ones in 217.

hundreds	tens	ones

2) Write how many hundreds, tens, and ones in 439.

____ hundreds ____ tens ____ ones

Write the number that is 100 more than 439.

____

3) Write the word name for 793.

_____

Write the word name for the number that is 10 less than 793.

_____

 Write a Story!

4) Write the number 962 in expanded form. Then write a story about 962 stickers.

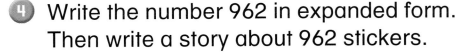

_____ + _____ + _____

_____

Chapter 21 Problem Solving Practice

four hundred nine **409**

Writing for Math

Write a story about the number of beads in the picture.

Use the words on the cards to help.

| hundreds | tens | ones |

Think

How many jars of 100 beads do I see in the picture? _____ jars

How many bracelets of 10 beads do I see in the picture?

_____ bracelets

How many other beads do I see in the picture? _____

_____ hundreds _____ tens _____ ones = _____

Solve

I can write my story now.

_____

_____

_____

Explain

I can tell you how my number matches my story.

_____

_____

_____

Chapter 21
Review/Test

Name_____

Write how many tens and how many ones.
Then write the number.

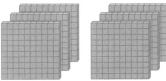

6 hundreds = _____ tens = _____ ones = _____

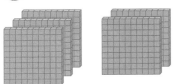

5 hundreds = _____ tens = _____ ones = _____

Write how many hundreds, tens, and ones.
Then write the number.

3 2 hundreds 8 tens 2 ones

hundreds	tens	ones

_____ + _____ + _____

_____

4 4 hundreds 0 tens 8 ones

hundreds	tens	ones

_____ + _____ + _____

_____

5 Jack has 200 blue marbles, 315 gold marbles, and 18 pink marbles. If Terry gives him 100 more blue marbles, how many blue marbles will he have in all?

_____ blue marbles

Spiral Review and Test Prep
Chapters 1–21

Choose the best answer.

1 Which figure is a pyramid?

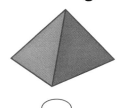 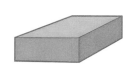 
◯ ◯ ◯ ◯

2 Miles had $1.00. He finds one coin. Now he has $1.25. Which coin did he find?

◯ ◯ ◯ ◯

3 Which number shows 5 hundreds, 2 tens, and 3 ones?

523   520   325   235
◯   ◯   ◯   ◯

Write the correct answer.

4 How many centimeters long is the string of beads?

_____ centimeters

5 What is the time?

_____

6 What could be the next number in this pattern? Explain how you know.

10, 20, 30, 40, 50, ____

_____

LOG ON www.mmhmath.com
For more Review and Test Prep

UNIT 6 CHAPTER 22

Number Relationships and Patterns

Pattern Song

SING TOGETHER

Sung to the tune of "Are You Sleeping?"

Find the pattern,

Find the pattern.

What comes next?

What comes next?

Should I pick a blue one?

Do I need a green one?

Look and see.

Look and see.

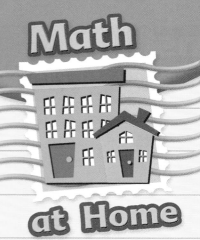

Math at Home

Dear Family,

I will learn about comparing and ordering 3-digit numbers in Chapter 22. Here are my math words and an activity that we can do together.

Love, _____

My Math Words

is greater than > :
27 > 26

is less than < :
26 < 27

is equal to = :
26 = 26

Home Activity

Write a few 2-digit numbers that are not in order. Have your child write the numbers from least to greatest.

Repeat the activity with other sets of numbers.

Books to Read

Look for these books at your local library and use them to help your child learn number relationships and patterns.

- **Benjamin's 365 Birthdays** by Judi Barrett and Ron Barrett, Atheneum Books, 1974.
- **How Much, How Many, How Far, How Heavy, How Long, How Tall Is 1000?** by Helen Nolan, Kids Can Press, 1995.
- **What's Next, Nina?** by Sue Kassirer, The Kane Press, 2001.

www.mmhmath.com
For Real World Math Activities

Name_____   **Compare Numbers**   ALGEBRA

Learn You can compare numbers using >, <, or =.

Math Words
is greater than >
is less than <
is equal to =

Compare 324 and 133.

First compare the hundreds.

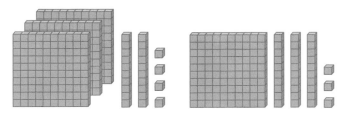

324 is greater than 133.

324 > 133

Compare 213 and 231.

The hundreds are the same. Compare the tens.

213 is less than 231.

213 < 231

Try It Compare. Write >, <, or =.

1. 347 > 197
2. 415 ◯ 429
3. 621 ◯ 635

4. 320 ◯ 352
5. 850 ◯ 750
6. 583 ◯ 540

7. 958 ◯ 779
8. 207 ◯ 210
9. 295 ◯ 232

10. 853 ◯ 853
11. 923 ◯ 927
12. 567 ◯ 567

13. **Write About It!** Explain how you would compare 157 and 162.

Chapter 22 Lesson 1   four hundred fifteen **415**

Practice Write >, <, or =.
Compare 128 and 124.

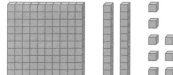  

128 is greater than 124.

128 > 124

The hundreds are the same. The tens are the same. Compare the ones.

14. 689 > 627
15. 372 ◯ 374
16. 450 ◯ 425

17. 281 ◯ 182
18. 105 ◯ 105
19. 789 ◯ 799

20. 601 ◯ 601
21. 233 ◯ 230
22. 955 ◯ 955

23. 723 ◯ 723
24. 325 ◯ 300
25. 252 ◯ 251

26. 533 ◯ 515
27. 142 ◯ 180
28. 697 ◯ 655

Problem Solving — **Logical Reasoning**

29. I am greater than 3 hundreds 2 tens and 2 ones. I am less than 3 hundreds 2 tens and 4 ones. What number am I?

    _____

30. I am greater than 8 hundreds 7 tens and 5 ones. I am less than 8 hundreds 7 tens and 7 ones. What number am I?

    _____

Math at Home: Your child learned how to compare 3-digit numbers.
Activity: Ask your child to name two numbers that are greater than 286 and then two numbers that are less than 550.

Name_____

Order Numbers on a Number Line

Learn You can use a number line to put numbers in order.

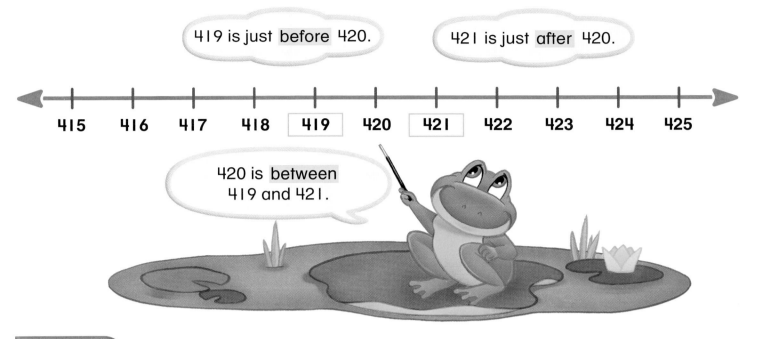

419 is just before 420.

421 is just after 420.

420 is between 419 and 421.

Try It Write the number that is just before.

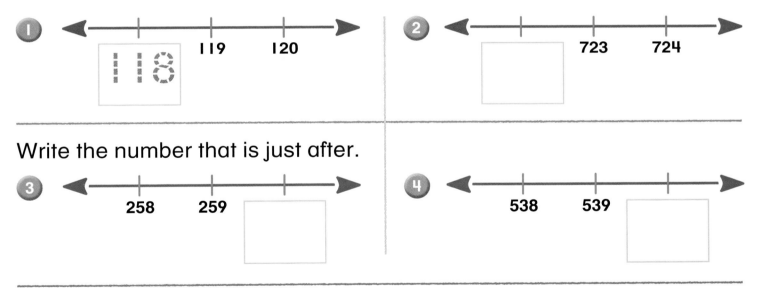

1. 118, 119, 120
2. ___, 723, 724

Write the number that is just after.

3. 258, 259, ___
4. 538, 539, ___

Write the number that is between.

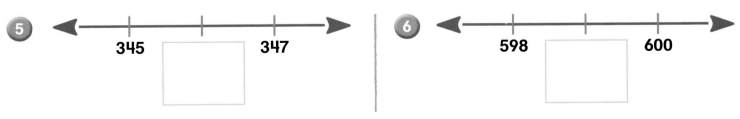

5. 345, ___, 347
6. 598, ___, 600

7. **Write About It!** How would you find the number that is just before 900?

Chapter 22 Lesson 2   four hundred seventeen **417**

Practice Write the numbers that come just before and just after.

8) 348 349 350

9) ___ 752 ___

10) ___ 820 ___

11) ___ 528 ___

12) ___ 903 ___

13) ___ 100 ___

Algebra • Patterns

Complete each pattern.

14) What are the next 3 shapes?

▲ ■ ● ▲ ■ ● ▲ ■ ● ___ ___ ___

This is a repeating pattern.

15) What is the next shape?

□   □□   □□□   □□□□   ___

This is a growing pattern.

Math at Home: Your child ordered numbers.
Activity: Pick a number from 100 to 999. Have your child name the numbers that come just before and just after the number.

418 four hundred eighteen

Name_____   **Order Numbers**

Learn You can order numbers by comparing. Order the numbers from least to greatest.

| 456 | 237 | 451 |

First compare the hundreds.
456   **2**37   **4**51

237 is the least.

456 and 451 have the same hundreds, so compare the tens.
237   4**5**6   4**5**1

The hundreds and tens are the same.

Since the hundreds and tens are the same compare the ones.
237   45**1**   45**6**

456 is greater than 451.

237 , 451 , 456

Try It Order the numbers from least to greatest.

1. | 592 | 425 | 530 | 425 , 530 , 592

2. | 765 | 762 | 318 | ____ , ____ , ____

Order the numbers from greatest to least.

3. | 219 | 481 | 247 | 481 , 247 , 219

4. | 703 | 726 | 717 | ____ , ____ , ____

5. **Write About It!** What is the order of the place value you need to look at to order numbers?

Chapter 22   Lesson 3   four hundred nineteen **419**

Practice Order the numbers from least to greatest.

6) 172 236 242 221   172, 221, 236, 242

7) 327 518 354 569   ____, ____, ____, ____

8) 436 431 620 400   ____, ____, ____, ____

9) 721 514 726 743   ____, ____, ____, ____

Order the numbers from greatest to least.

10) 523 357 525 341   525, 523, 357, 341

11) 619 630 647 632   ____, ____, ____, ____

12) 871 287 284 835   ____, ____, ____, ____

13) 901 918 908 981   ____, ____, ____, ____

Make it Right

14) Petro put numbers in order from greatest to least.

528, 438, 440

Why is Petro wrong? Make it right.

420 four hundred twenty

Name _____   **Number Patterns**   ALGEBRA

Learn You can use number patterns to help you count.

Count by hundreds. Each number is 100 more.

| 150 | 250 | 350 | 450 | 550 | 650 | 750 |

Count by tens. Each number is 10 more.

| 113 | 123 | 133 | 143 | 153 | 163 | 173 |

Count by ones. Each number is 1 more.

| 321 | 322 | 323 | 324 | 325 | 326 | 327 |

Try It Write the missing number. Then circle the counting pattern.

Numbers	Pattern: Count by
① 140, 150, 160, __170__, 180	hundreds (tens) ones
② 365, 465, _____, 665, 765	hundreds tens ones
③ _____, 235, 236, 237, 238	hundreds tens ones
④ 528, 538, 548, _____, 568	hundreds tens ones

⑤ **Write About It!** How can you tell if a number pattern is counting by hundreds?

Chapter 22   Lesson 4   four hundred twenty-one   **421**

Practice Write the missing number. Then circle the counting pattern.

Count by tens.

428, 438, 448, 458, 468

Numbers	Pattern: Count by		
6) 920, 930, __940__, 950, 960	hundreds	(tens)	ones
7) 432, 532, 632, _____, 832	hundreds	tens	ones
8) 785, _____, 787, 788, 789	hundreds	tens	ones
9) 142, 143, _____, 145, 146	hundreds	tens	ones
10) 500, 510, 520, _____, 540	hundreds	tens	ones
11) 299, 399, 499, 599, _____	hundreds	tens	ones

Problem Solving **Critical Thinking**

12) The art store orders 10 brushes each month. How many brushes does the shop order in 5 months?

_____ brushes

Number of Months	Brushes
1	10
2	20
3	30
4	40
5	?

Math at Home: Your child described and extended number patterns.
Activity: Pick a number between 100 and 500. Have your child count by ones, tens, or hundreds.

Name_____

Count Forward, Count Backward

Learn You can use a number line to count forward or backward.

Count forward by tens.

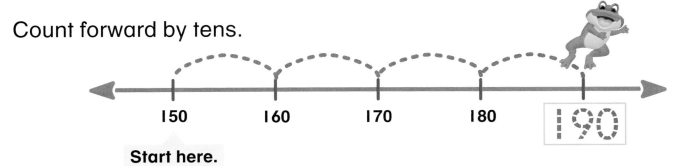

Start here.

Count backward by tens.

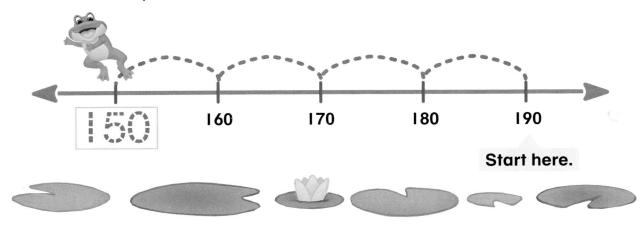

Start here.

Try It Count forward or backward.

Count forward by tens.	Count backward by tens.
1. 256, __266__, __276__	5. 371, __361__, __351__
2. 827, ____, ____	6. 594, ____, ____
3. 479, ____, ____	7. 643, ____, ____
4. 132, ____, ____	8. 238, ____, ____

9. **Write About It!** What number would you get if you count forward by ten from 299? Explain.

Chapter 22 Lesson 5   four hundred twenty-three **423**

Practice Count forward or backward by ones.

Count forward by ones.
10. 347, __348__, __349__
11. 862, _____, _____
12. 648, _____, _____
13. 599, _____, _____

Count backward by ones.
14. 105, __104__, __103__
15. 424, _____, _____
16. 200, _____, _____
17. 701, _____, _____

Count forward or backward by hundreds.

Count forward by hundreds.
18. 238, __338__, __438__
19. 417, _____, _____
20. 672, _____, _____
21. 800, _____, _____

Count backward by hundreds.
22. 921, __821__, __721__
23. 678, _____, _____
24. 459, _____, _____
25. 503, _____, _____

Problem Solving Number Sense

26. Look at the number. Complete the charts.

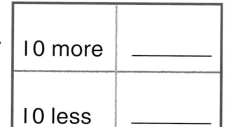

10 more	_____
10 less	_____

100 more	_____
100 less	_____

Math at Home: Your child learned to count forward and backward.
Activity: Pick a 3-digit number. Ask your child to count forward and backward by tens. Continue with other numbers.

424 four hundred twenty-four

Name _____

Read Plan Solve Look Back

Problem Solving Strategy

Make a Table • Algebra

Sometimes you can make a table to help solve a problem.

Briana and Amber want to make 5 bracelets. They need 10 beads for each bracelet. How many beads do they need in all?

Read

What do I already know? _____ bracelets

_____ beads for each bracelet

What do I need to find? _____

Plan

I can make a table. Then look for the pattern in the table.

Solve

I can carry out my plan.

The girls need ___50___ beads.

Number of Bracelets	1	2	3	4	5
Number of Beads	10	20			

Look Back

How can I check my answer? _____

_____

Chapter 22 Lesson 6

four hundred twenty-five **425**

Read > Plan > Solve > Look Back

Complete the table to solve.

1. Jessica puts a dime in her piggy bank every week. How much can she save in 5 weeks?

   _____ ¢

Weeks	1	2	3	4	5
Money	10¢				

2. David collects toy cars. He can fit 6 toy cars on a shelf. How many toy cars can he fit on 5 shelves?

   _____ toy cars

Shelves	1	2	3	4	5
Cars					

3. Joe's class uses 1 box of paper every week. There are 100 sheets of paper in a box. How many sheets of paper do they use in 5 weeks?

   _____ sheets

Weeks	1	2	3	4	5
Sheets					

4. 50 mugs are sold at the craft fair every day. How many mugs are sold in 5 days?

   _____ mugs

Days	1	2	3	4	5
Mugs					

Math at Home: Your child learned to solve problems by making a table.
Activity: Make a chart for a problem such as, "Jan's family makes a road trip. They drive 100 miles a day. How far do they travel in 5 days?" Help your child fill in the chart to solve.

Game Zone

Practice at School ★ Practice at Home

Name_____

Trading Tens

▶ Choose a workmat. Take turns.

▶ Toss the 🎲. Put that many tens on your workmat.

▶ When you have 10 ▬▬▬, trade them for 1 ▪.

▶ The first player to get 5 ▪ wins.

 2 players

You Will Need

🎲

20 ▬▬▬

9 ▪

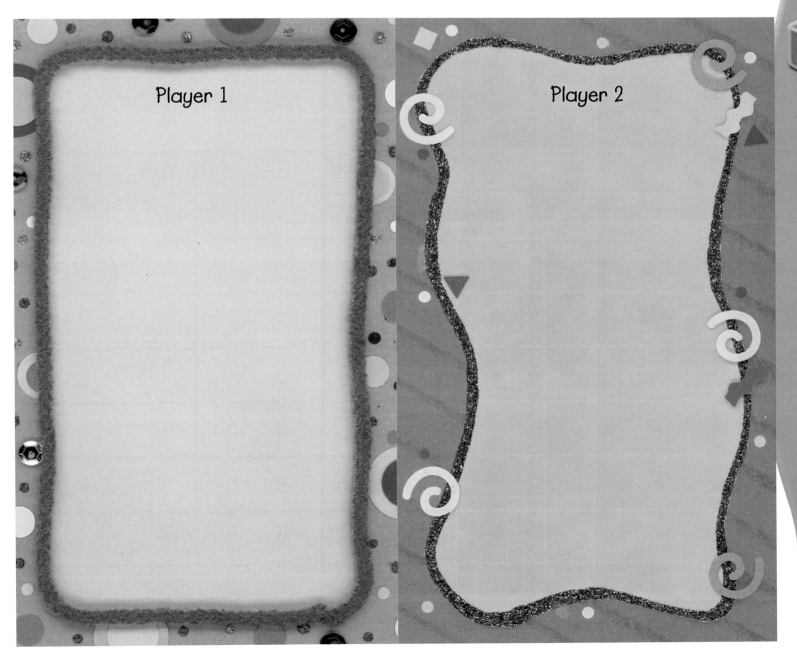

Player 1

Player 2

Chapter 22 Game Zone

four hundred twenty-seven **427**

Technology Link

Place Value • Computer

Use to make numbers.

Choose a mat to show place value.

Stamp out 3 ▇.

Stamp out 5 ▭.

Stamp out 9 ▪.

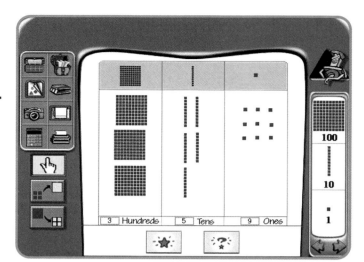

What is the number? __359__

You can use the computer to stamp out hundreds, tens, and ones.

Then write the number.

Stamp Out			What is the Number?
▇	▭	▪	
2	6	1	261
5	0	2	
7	8	3	
6	2	0	
1	5	8	
4	3	4	

 For more practice use Math Traveler.™

Chapter 22
Review/Test

Name_____

Write the missing numbers.

① 645 _____ 647   ② 398 _____ 400

Compare. Write >, <, or =.

③ 545 ◯ 645   ④ 234 ◯ 234   ⑤ 333 ◯ 233

⑥ Write the numbers in order from greatest to least.

645 219 637 302   _____, _____, _____, _____

⑦ Write the numbers in order from least to greatest.

812 218 182 821   _____, _____, _____, _____

Write the missing number in each pattern.
Circle the pattern you used to count.

⑧ 225, 325, 425, _____, 625, 725

hundreds   tens   ones

⑨ 434, _____, 454, 464, 474, 484

hundreds   tens   ones

⑩ June's family takes a trip. They drive 100 miles a day. How far do they travel in 5 days?

_____ miles

Days	1	2	3	4	5
Miles					

Chapter 22 Review/Test   four hundred twenty-nine **429**

Spiral Review and Test Prep
Chapters 1–22

Choose the best answer.

1. What is the time?

    4:45   4:15   3:45   3:15

2. Which figure is a cylinder?

3. Hector has 27 stickers.
   Kyra gives him 18 more stickers.
   Which shows how many stickers Hector has now?

   27 + 18   27 − 18   45 + 27   45 − 27

Solve.

4. Britney had $1.40. Then she found 3 nickels.
   Now how much money does Britney have? $_____._____

5. Make a skip-counting pattern. Tell about your pattern.

_____

www.mmhmath.com
For more Review and Test Prep

430 four hundred thirty

UNIT 6 CHAPTER 23

3-Digit Addition

Main Attraction

Everybody step right up—

Our show's a special treat!

A hundred acts, a million thrills—

So hurry, take your seat!

And in this ring, a superstar

As big as big can be.

Its feet are size six hundred ten,

And mine are just size three!

Math at Home

Dear Family,

I will learn how to add 3-digit numbers in Chapter 23. Here are my math words and an activity that we can do together.

Love, _____

My Math Words

addend:

hundreds:

2 3 4
↑
2 hundreds

regroup:

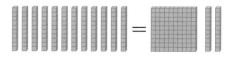

12 tens = 1 hundred 2 tens

Home Activity

Make a hundreds, tens and ones chart.

hundreds	tens	ones

Say a number between 100 and 999, such as 154, 308, or 620.

Have your child write the number as hundreds, tens, and ones in the chart.

Repeat the activity several times.

Books to Read

Look for these books at your local library and use them to help your child add 3-digit numbers.

- **The 512 Ants on Sullivan Street** by Carol A. Losi, Scholastic, 1997.
- **The Case of the Shrunken Allowance** by Joanne Rocklin, Scholastic, 1998.

www.mmhmath.com
For Real World Math Activities

432 four hundred thirty-two

Name_____

Add Hundreds

Learn You can use addition facts to add hundreds.
Add 500 + 300.

*I know 5 + 3 = 8.
I use that to add hundreds.
500 + 300 = 800*

5 hundreds + 3 hundreds = __8__ hundreds

500 + 300 = __800__

Try It  Add.

① 300 + 100 = __400__

② 200 + 200 = ____

③ 100 + 400 = ____

④ 300 + 600 = ____

⑤ 200 + 500 = ____

⑥ 400 + 500 = ____

⑦ 700 + 100 = ____

⑧ 200 + 400 = ____

⑨ **Write About It!** What fact can you use to add 400 + 300?

Practice Add.

I know 4 + 2 = 6.
I use that to add 400 + 200.

10. 400 + 200 = 600

11. 300 + 400 = ___

12. 800 + 100 = ___

13. 700 + 200 = ___

14. 300
    +300

15. 500
    +300

16. 500
    +100

17. 300
    +200

18. 100
    +300

19. 200
    +600

20. 400
    +400

21. 700
    +200

22. 200
    +100

23. 300
    +400

24. 100
    +800

25. 100
    +600

Math at Home: Your child used basic facts to add hundreds.
Activity: Have your child show you how to add 200 + 100.

Extra Practice

Name _____

Add. Color exercises where you regroup 🖍.
Color other exercises 🖍.

$35¢$
$+51¢$
$86¢$

¢ is a cent sign.

$54¢$
$+8¢$

$16¢$
$+27¢$

$53¢$
$+44¢$

$28¢$
$+53¢$

$40¢$
$+39¢$

$20¢$
$+74¢$

$42¢$
$+38¢$

$67¢$
$+8¢$

$42¢$
$+25¢$

$15¢$
$+27¢$

$58¢$
$+19¢$

$36¢$
$+15¢$

$39¢$
$+56¢$

$30¢$
$+61¢$

Chapter 23 Extra Practice

four hundred thirty-five **435**

Extra Practice

Use the clues.
Draw the hands on the clock.
Write the time.

1. My minute hand is on 12.
   My hour hand is on 9.
   What time am I?

     9:00

2. My minute hand is on 6.
   My hour hand is between 3 and 4.
   What time am I?

3. My minute hand is on 3.
   My hour hand is between 7 and 8.
   What time am I?

4. My minute hand is on 9.
   My hour hand is between 10 and 11.
   What time am I?

    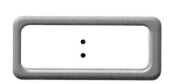

5. I am two hours after 1 o'clock.
   What time am I?

6. **Write About It!** The short hand is on the 2. The long hand is on the 6. What time is it?

LOG ON wwww.mmhmath.com
For more Practice

Math at Home: Your child practiced telling time.
Activity: Have your child tell you where the clock hands are at 8:30. Repeat using other times.

Name_____

Problem Solving Skill
Reading for Math

The Circus

The circus is in town. 175 people come to the show on Monday, 115 on Tuesday, and 104 on Wednesday. On Thursday there's no show. But 178 people come on Friday. And over the weekend, 694 people in all come to see the circus show!

Reading Skill

Make Inferences

1. Why do you think more people came to the show on the weekend?

   _____

   _____

2. How many people in all saw the circus show on Monday and Tuesday? _____ people

3. How many people in all saw the circus show on Wednesday and Friday? _____ people

4. How many people came to the show on Tuesday and Friday?

   _____ people

Chapter 23   Lesson 4

four hundred forty-one **441**

Sunday at the Circus

Many people come to the circus on Sunday. They buy 167 bags of peanuts and 215 bags of popcorn. They buy 383 bottles of water and 231 bottles of soda.

 Make Inferences

5. Why do you think people buy so much water or soda to drink?

   _____

   _____

6. How many bottles of water and soda do they buy? _____ bottles

7. How many bags of peanuts and popcorn do they buy? _____ bags

 Math at Home: Your child made inferences to answer questions.
Activity: Have your child make an inference or judgment about how many people were at the circus on Sunday.

Name _____

Problem Solving Practice

Solve.

1. Yesterday the 🐘 ate 213 peanuts.

   Today the 🐘 ate 348 peanuts.

   How many 🥜 did it eat in all?

   213 + 348 = _____

2. Last year the 🎪 put on 290 shows.

   This year the 🎪 put on 313 shows.

   How many shows did the 🎪 put on in all?

   290 + 313 = _____

Write a Story!

3. Use the number sentence to write an addition problem about a circus. Find the sum.

   318 + 269 = _____

   _____

   _____

4. Last weekend the circus sold 342 posters. This weekend the circus sold 386 posters. How many posters did they sell in all?

   342 + 386 = _____

5. Last week the circus traveled 266 miles to get to our town. This week it traveled 314 miles to the next city. How many miles did it travel in all?

   266 + 314 = _____

Writing for Math

Is Eric's estimate reasonable?

108 + 294 is about 400.

Think

I can find the nearest hundred for each number.

Solve

I know that 108 is between 100 and 200. It is closer to 100.

I know that 294 is between 200 and 300. It is closer to 300.

Then I estimate.

____ + ____ = ____

Explain

I can tell you why the answer is reasonable.

_____

_____

_____

_____

Chapter 23
Review/Test

Name_____

Add.

1. $200 + 400 = $ _____

2. $100 + 700 = $ _____

3. $300 + 600 = $ _____

4. $500 + 200 = $ _____

Add. Regroup when you need to.

5.
hundreds	tens	ones
	☐	
1	4	9
+1	2	3

6.
hundreds	tens	ones
	☐	
2	7	1
+1	0	4

7.
hundreds	tens	ones
☐		
3	8	2
+1	7	5

8.
hundreds	tens	ones
☐		
3	7	6
+2	4	3

9. The elephants ate 150 peanuts for breakfast. They ate 250 peanuts for lunch. How many peanuts did they eat in all? _____

10. The big elephant eats 400 peanuts for dinner. The little elephant eats 150 peanuts for dinner. Why do you think the little elephant eats fewer peanuts?

_____

Spiral Review and Test Prep
Chapters 1–23

1. Choose the number for: 3 hundreds 6 ones.

   36 ○   306 ○   360 ○   3006 ○

2. Which is 100 more than 428?

   328 ○   429 ○   438 ○   528 ○

Use the table.

Bags of Snacks Sold		
Day	Popcorn	Peanuts
Saturday	245	176
Sunday	324	118

3. What was the total number of bags of peanuts sold on Saturday and Sunday? _____ bags

4. How many bags of popcorn and peanuts did the circus sell on Sunday? _____ bags

5. Use the numbers 2, 7, and 9. What is the greatest 3-digit number you can write? _____

   Explain your answer. _____

   _____

   _____

LOG ON www.mmhmath.com
For more Review and Test Prep

UNIT 6 CHAPTER 24

3-Digit Subtraction

SEA SONG

Sung to the tune of "My Bonnie Lies Over the Ocean"

I went to the beach on vacation,

To take a cool dip in the sea,

And when I went into the water,

A large school of fish swam by me.

I stopped and I started to count them—

I got up to 153.

But just then a wave knocked me over,

And swept all the fish out to sea!

Math at Home

Dear Family,

I will learn how to subtract 3-digit numbers in Chapter 24. Here are my math words and an activity that we can do together.

Love, _____

My Math Words

difference:

← difference

regroup:

1 hundred 1 ten = 11 tens

Home Activity

Work with your child to make up several subtraction problems involving 1- and 2-digit numbers. For example, subtract the number of letters in your name from 43.

Have your child solve the problems.

Books to Read

In addition to this library book, look for the Time For Kids math story that your child will bring home at the end of this unit.

- **Tightwad Tod** by Daphne Skinner, The Kane Press, 2001.
- **Time For Kids**

www.mmhmath.com
For Real World Math Activities

Name_____

Subtract Hundreds

Learn You can use subtraction facts to subtract hundreds.

Subtract 600 − 200.

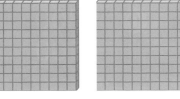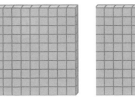
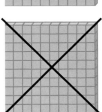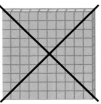

I know that 6 − 2 = 4.
I use that to subtract hundreds.
600 − 200 = 400

6 hundreds − 2 hundreds = __4__ hundreds

600 − 200 = __400__

Try It Subtract.

① 500 − 100 = __400__

② 900 − 200 = ____

③ 700 − 400 = ____

④ 600 − 100 = ____

⑤ 800 − 500 = ____

⑥ 400 − 300 = ____

⑦ 300 − 100 = ____

⑧ 200 − 100 = ____

⑨ **Write About It!** What fact can you use to subtract 900 − 300?

Chapter 24   Lesson 1

four hundred forty-nine **449**

Practice Subtract.

10) 700 − 300 = **400**

Think: 7 − 3 = 4. So 700 − 300 = 400.

11) 800 − 100

12) 400 − 100

13) 800 − 400

14) 500 − 200

15) 600 − 400

16) 900 − 300

17) 400 − 200

18) 300 − 200

19) 600 − 300

20) 700 − 200

21) 500 − 400

22) 800 − 300

23) 900 − 100

24) 500 − 300

25) 600 − 500

26) 700 − 100

27) 800 − 600

Spiral Review and Test Prep

Choose the best answer.

28) Which number is 100 less than 400?

   ○ 300   ○ 400   ○ 500   ○ 600

29) Which would be the best unit to use to measure the length of a pencil?

   ○ inch   ○ foot   ○ yard   ○ meter

 Math at Home: Your child used basic facts to subtract hundreds.
Activity: Have your child subtract 900 − 700.

Name _____

Regroup Tens as Ones

Learn

You can use hundreds, tens, and ones models to subtract. Use a workmat and 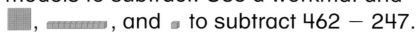 to subtract 462 − 247.

Step 1

To subtract the ones you need to take away 7. You need to regroup 1 ten as 10 ones. Subtract the ones.

hundreds	tens	ones
	[5]	[12]
4	6̸	2̸
− 2	4	7
		5

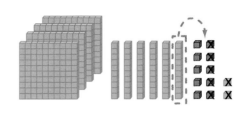

Step 2

To subtract the tens take away 4. Write how many tens are left.

hundreds	tens	ones
	[5]	[12]
4	6̸	2̸
− 2	4	7
	1	5

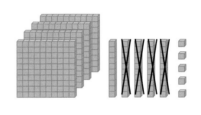

Step 3

To subtract the hundreds take away 2. Write how many hundreds are left.

hundreds	tens	ones
	[5]	[12]
4	6̸	2̸
− 2	4	7
2	1	5

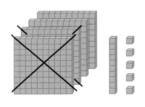

Your Turn

Use  to subtract.

1.

hundreds	tens	ones
	[6]	[14]
3	7̸	4̸
− 1	4	8
2	2	6

hundreds	tens	ones
	☐	☐
2	8	3
− 1	3	5

hundreds	tens	ones
	☐	☐
3	5	1
− 2	3	7

2. **Write About It!** How is subtracting 3-digit numbers like subtracting 2-digit numbers?

Chapter 24 Lesson 2 — four hundred fifty-three **453**

Practice Use ▮, ▬, and ▪ to subtract.

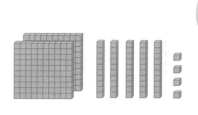

If there are not enough ones to subtract you need to regroup.

3)
hundreds	tens	ones
2	⁴5	¹⁴4
−1	3	9
		5

4)
hundreds	tens	ones
4	8	5
−1	5	7

hundreds	tens	ones
5	9	6
−3	4	8

hundreds	tens	ones
3	4	7
−2	1	3

5)
hundreds	tens	ones
2	6	3
−1	4	6

hundreds	tens	ones
6	7	4
−2	2	5

hundreds	tens	ones
4	5	1
−3	1	4

Algebra • Missing Numbers

6) ▢▢
 −3 8
 ‾‾‾‾
 1 5

7) 6 1
 −▢ ▢
 ‾‾‾‾
    2 3

8) ▢▢
 −3 8
 ‾‾‾‾
    9

Math at Home: Your child subtracted 3-digit numbers by regrouping tens.
Activity: Have your child draw models to show you how to subtract 284 − 136.

Chapter 23 Lesson 2

Regroup Hundreds as Tens

HANDS ON Activity

Use workmat and ▢, ▭, and ▫ to find the difference of 528 and 176.

Step 1
To subtract the ones take away 6 ▫.
Write how many ones are left.

hundreds	tens	ones
5	2	8
− 1	7	6
		2

Step 2
To subtract the tens you need to take away 7 ▭.
You need to regroup 1 hundred as 10 tens.
Now subtract the tens.

hundreds	tens	ones
4	12	
5̸	2̸	8
− 1	7	6
	5	2

Step 3
Subtract the hundreds.
Write how many hundreds are left.

hundreds	tens	ones
4	12	
5̸	2̸	8
− 1	7	6
3	5	2

Your Turn Use ▢, ▭, and ▫ to subtract.

1.

hundreds	tens	ones
2	11	
3̸	1̸	5
− 1	5	2
1	6	3

hundreds	tens	ones
4	5	7
− 1	8	5

hundreds	tens	ones
5	3	4
− 3	7	1

2. ✏ **Write About It!** Explain how you know when to regroup hundreds to subtract.

Chapter 24 Lesson 3   four hundred fifty-five **455**

Practice Use ▦, ▭, and ▫ to subtract.

If there are not enough tens to subtract you need to regroup.

3.
hundreds	tens	ones
²3̷	¹⁴4̷	9
− 1	7	5
1	7	4

4.
hundreds	tens	ones
☐	☐	
3	2	6
− 1	4	5

hundreds	tens	ones
☐	☐	
5	3	7
− 2	9	4

hundreds	tens	ones
☐	☐	
6	5	4
− 1	8	4

5.
hundreds	tens	ones
☐	☐	
4	9	7
− 2	6	5

hundreds	tens	ones
☐	☐	
7	1	8
− 3	4	2

hundreds	tens	ones
☐	☐	
5	4	6
− 1	9	3

Make it Right

6.
hundreds	tens	ones
☐	☐	
4	3	8
− 1	7	5
3	4	3

This is how Ben subtracted 438 − 175. Why is Ben wrong? Make it right.

Math at Home: Your child subtracted 3-digit numbers by regrouping hundreds.
Activity: Have your child show you how to subtract 438 − 251 and explain the regrouping to you.

Name _____

Problem Solving Strategy

Work Backward

Sometimes you start with what you know and work backward to solve a problem.

The Wilson family is taking a 350 mile road trip. They have 218 miles left to go. How many miles did they already travel?

Read

What do I already know?   The trip is _____ miles long.

They have _____ miles left to go.

What do I need to find out?   _____

Plan

I can subtract the miles they have left to go. That will tell me how many miles they have already traveled.

Solve

I can carry out my plan.

The family has already traveled _____ miles.

$$\begin{array}{r} \overset{4\ 10}{3\cancel{5}\cancel{0}} \\ -218 \\ \hline 132 \end{array}$$

Look Back

How can I check my answer? _____

_____

Chapter 24   Lesson 5

four hundred fifty-nine   **459**

Read · Plan · Solve · Look Back

Solve.

Draw or write to explain.

1. There are 293 pages in a book. Liz has 148 pages left to read. How many pages did Liz read already?

_____ pages

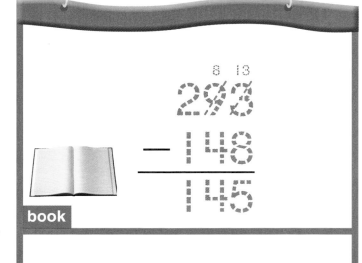
book

2. Mr. Davis opens a new box of straws. The children use 116 straws. There are 134 straws left in the box. How many straws were in the new box?

_____ straws

straw

3. There are 375 sheets of paper. The class uses some paper for a project. Now there are 257 sheets of paper left. How much paper did the class use for the project?

_____ sheets of paper

sheets of paper

4. Brenda bought some stickers for $2.48. Now she has $3.27. How much money did Brenda have to begin with?

_____

stickers

Math at Home: Your child learned to solve problems by working backward.
Activity: Have your child tell you how he or she solved exercise 4.

Name _____

Game Zone
Practice at School ★ Practice at Home

It All Adds Up!

- Work with a partner. Choose one of the problems below. Roll the number cube 6 times.
- Put each number in one place in the addition problem.
- Find the sum.
- The player with the answer closest to 500 wins.

👥 2 players

You Will Need
Number Cube

Player 1		
hundreds	tens	ones
☐	☐	☐
+ ☐	☐	☐

Player 2		
hundreds	tens	ones
☐	☐	☐
+ ☐	☐	☐

Technology Link

Dollars and Cents • Calculator

Use a to find the new price of the book.

You Will Use

> You need to use the decimal point key for amounts of money.

Show $5.29 on the calculator.

Press   .

You see .

Then subtract $1.50.

Press   . You see 3.79.

You can use addition to check the answer.

Find $3.79 + $1.50.

Press   5.29.

Use a to solve. Check your answer.

1 Steve wants to buy a baseball cap for $6.75 and a baseball for $2.89. How much money does he need in all?

_____

2 A picture frame was $9.65. The frame goes on sale for $1.75 less than the full price. How much does the frame cost?

_____

Chapter 24
Review/Test

Name _____

Subtract.

1) 900 − 400 = _____   2) 700 − 100 = _____

3) 800 − 200 = _____   4) 500 − 300 = _____

Subtract. You can use ▦, ▭, and ▪ to help.

5)
hundreds	tens	ones
□	□	□
5	7	8
− 1	4	3

6)
hundreds	tens	ones
□	□	□
4	9	2
− 2	3	6

Add or subtract. Estimate to check.

7) $3.79   8) $6.26
   − 1.04 − ____   + 2.80 + ____

Solve.

9) There are 283 pages in a book. Ella has 145 pages left to read. How many pages did Ella read already? _____

10) Jay bought a book for $4.20. Now he has $2.30. How much money did Jay have to begin with? _____

Chapter 24 Review/Test   four hundred sixty-three **463**

Spiral Review and Test Prep
Chapters 1–24

Choose the best answer.

1) In the number 345, the 3 means

- 3
- 30
- 300
- 400

2) What is the value of 400 + 30 + 6?

- 346
- 436
- 634
- 643

3) Which number shows 10 more than 278?

- 268
- 278
- 288
- 378

4) Complete.

5) The class collected 326 cans and 248 bottles. How many cans and bottles did the class collect?

_____ cans and bottles

6) Write a number that comes between 431 and 440. Tell how you know.

_____

_____

www.mmhmath.com
For more Review and Test Prep

TIME FOR KIDS

Name _____

Juanita has a collection of 101 tiny toys.

Vicki has 108 tiny toys.
Who has more tiny toys?

How many more?

_____

$$\begin{array}{r} 108 \\ -101 \\ \hline \end{array}$$

TIME FOR KIDS

Counting Collections

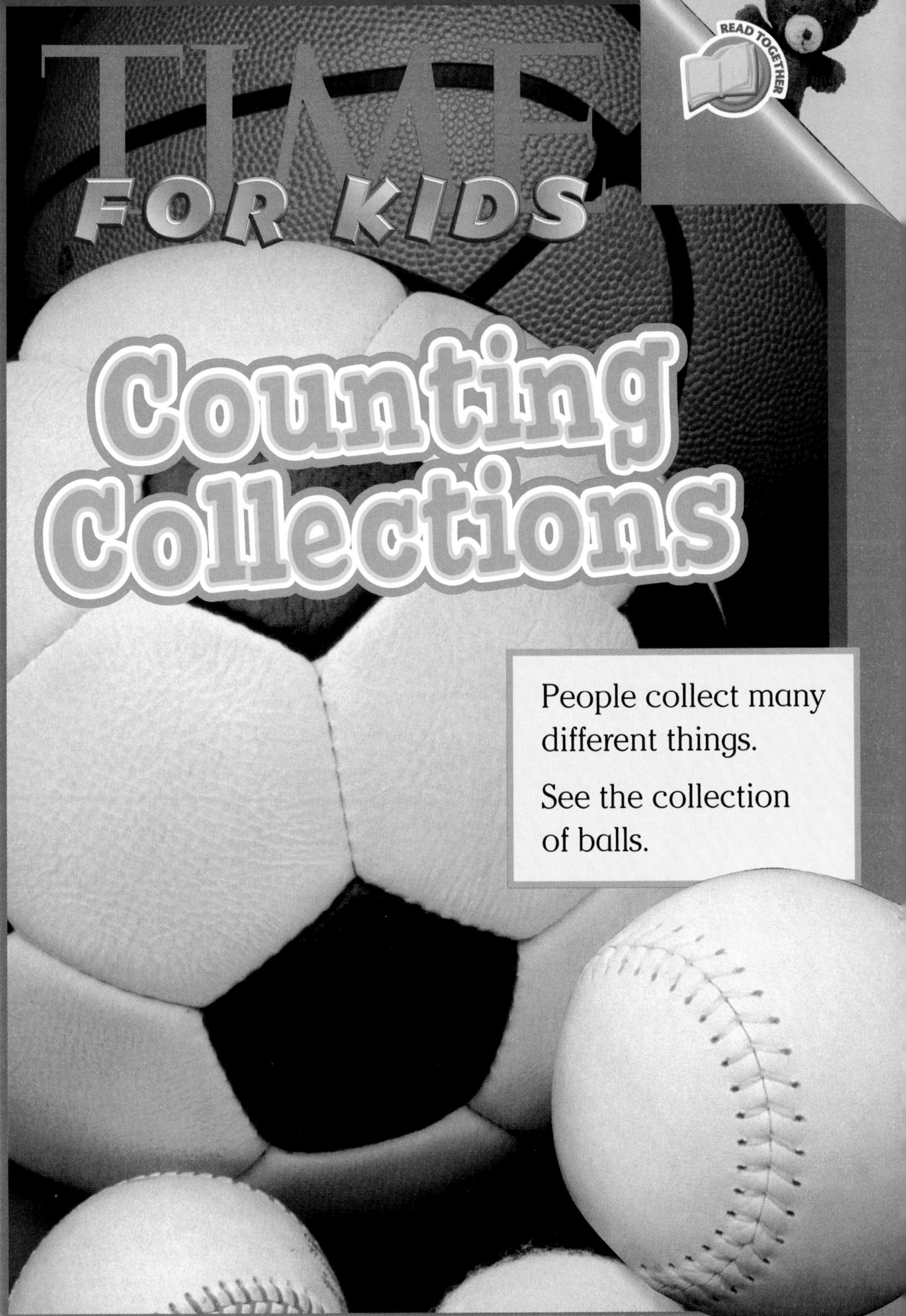

People collect many different things. See the collection of balls.

There are at least 125 stamps in this collection.

One jar has about 110 marbles.
There are about 250 coins in the other jar.

110 < 250

There are fewer marbles than coins.

You can count more than 100 seashells in this collection.

Name _____

Linking Math and Science

Addition and Water Use

Water comes from Earth. It is a natural resource we need to live. If we use too much water, we could run out of it. How could people save, or conserve, water?

Science Words
natural resource
conserve

Water is a natural resource we need.

A leaky faucet uses up to 15 gallons of water each day.

Circle the word or words to complete each sentence.

1. Things that come from the Earth are _____.

   natural resources   water

2. We must save or _____ water.

   use up   conserve

Unit 6 Linking Math and Science

What to Do

- Use  or ▨, ▭, ▪.
- Choose one water use from the chart.
- Find how many gallons of water you use in a day. Write the total.
- Find how many gallons of water you use in a week. Write the total.

Water Use	Gallons Used
Flush toilet	5 gallons
Wash hands	1 gallon
Take bath	35 gallons
Take shower	25 gallons

Water Use	
Number of times in 1 day	
Gallons in 1 day	
Gallons in 1 week	

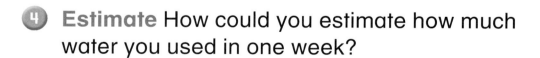

Use your data to solve.

3 **Measure** How did you find the amount of water you used in one day?

_____

4 **Estimate** How could you estimate how much water you used in one week?

_____

5 **Compare** How does the amount of water you use compare with the amount your classmates use?

_____

Math at Home: Your child applied addition to find how much water they use.
Activity: Repeat this activity. Choose a different water use from the chart. Have your child write the addition sentences for the daily use.

Name_____

Unit 6
Study Guide and Review

Math Words
Draw lines to match.

1. = — — — — — — is equal to
2. >   is greater than
3. <   is less than

Skills and Applications
Number Relationships and Patterns (pages 415-422)

Examples

Compare numbers.
Look at the hundreds first.
Then look at the tens and the ones.

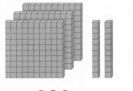

320   >   234
320 is greater than 234.

4. 821 ◯ 615
5. 652 ◯ 693
6. 901 ◯ 900
7. 215 ◯ 330

8. | 610 | | 630 |
   | 640 | 650 | |

Use number patterns to help you count.

Count by tens.

350, 360, 370, 380, 390

Count by hundreds.

418, 518, 618, 718, 818

9. | 299 | | 499 |
   | 599 | | 799 |

Unit 6   Study Guide and Review   four hundred sixty-seven   **467**

Skills and Applications

Add and Subtract 3-Digit Numbers (pages 432-444; 448-460)

Examples

Add. Regroup if necessary.

hundreds	tens	ones
[1]	[1]	
2	7	7
+ 2	3	5
5	1	2

10.

hundreds	tens	ones
☐	☐	
4	8	4
+ 3	6	7

Subtract. Regroup if necessary.

hundreds	tens	ones
☐	[6]	[15]
6	7̸	5̸
− 3	4	9
3	2	6

11.

hundreds	tens	ones
☐	☐	☐
4	2	8
− 2	8	6

Problem Solving — Strategy

(pages 421–426)

Use a table to solve.

12. The craft shop orders 200 boxes of beads each week. How many boxes of beads does the shop order in 4 weeks? 5 weeks?

Weeks	Boxes
1	200
2	400
3	600
4	_____
5	_____

Math at Home: Your child learned about number relationships and patterns to 1,000 and adding and subtracting 3-digit numbers.
Activity: Have your child use these pages to review number relationships and addition and subtraction of 3-digit numbers.

Name_____

Unit 6 Performance Assessment

Addition and Subtraction

Show as many combinations as you can.

335 695 529 205 148 179 285

| Find 2 numbers with a sum less than 850. Show the sums. | Find 2 numbers with a difference less than 200. Show the differences. |

_____   _____
_____   _____
_____   _____
_____   _____
_____   _____
_____   _____
_____   _____

 You may want to put this page in your portfolio.

e-Journal www.mmhmath.com
Write about math

Unit 6
Enrichment

Adding and Subtracting Money

Some friends went to the toy store.
Look at what they bought.
Then answer the questions.

A toy can be bought more than once.

1 Paul bought 2 items totaling $4.53.
Which 2 items did he buy?

_____

2 Megan bought 2 items totaling $3.86.
Which 2 items did she buy?

_____

3 Samuel paid $4.00 for 1 item. He got
$1.58 in change. Which item did he buy?

_____

4 Rachel paid $5.00 for 1 item. She got
$2.44 in change. Which item did she buy?

_____

UNIT 7

CHAPTER 25

Fractions

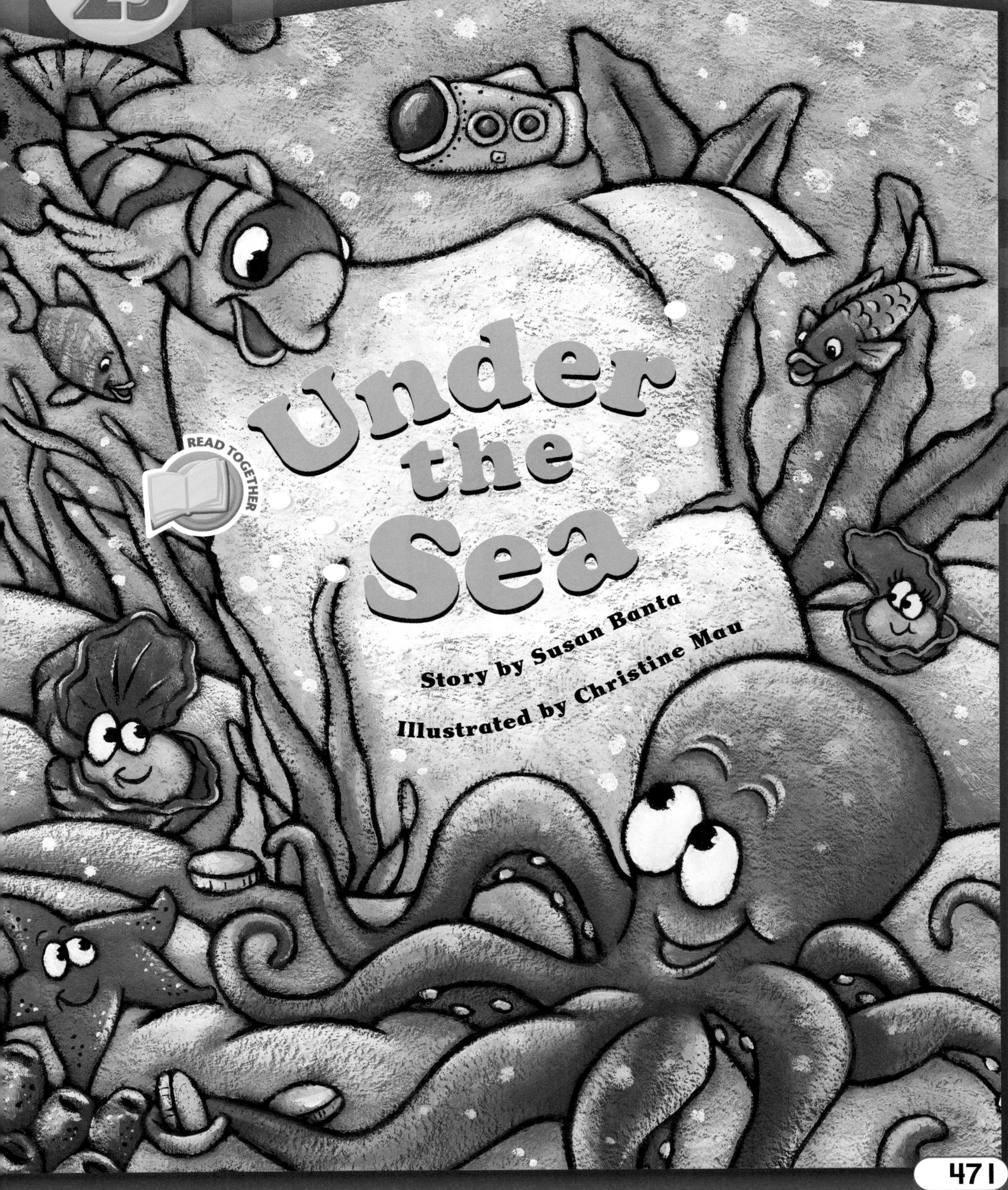

Under the Sea
READ TOGETHER
Story by Susan Banta
Illustrated by Christine Mau

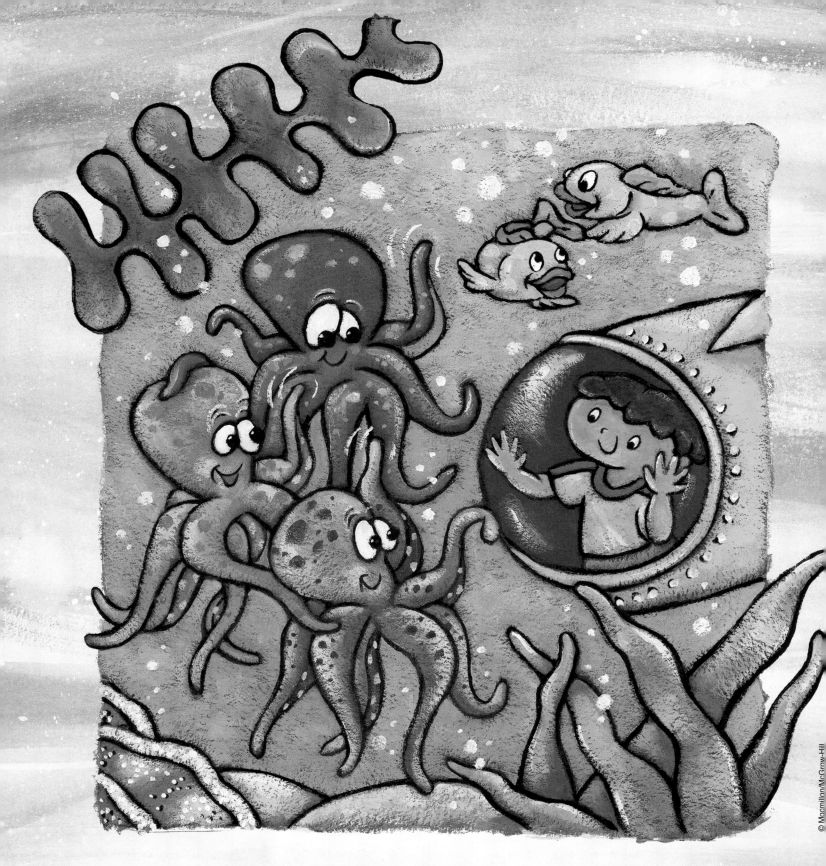

How many sea creatures swim with me?

_____ sea creatures

Only _____ of them are yellow fish.

How many starfish swim with me?

____ starfish

Only ____ starfish is hiding.

How many fish swim with me?

_____ fish

Only _____ of the fish are purple.

How many dolphins swim with me?

____ dolphins

____ of the dolphins are light gray.

Math at Home

Dear Family,

  I will learn about fractions in Chapter 25. Here are my math words and an activity that we can do together.

  Love, _____

My Math Words

fraction:
a number that names part of a whole or a group

$\frac{1}{8}$ $\frac{1}{6}$ $\frac{1}{4}$ $\frac{1}{3}$ $\frac{1}{2}$

is greater than (>):

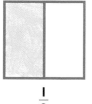

$\frac{1}{2}$ > $\frac{1}{4}$

is less than (<):

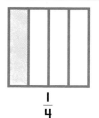 

$\frac{1}{4}$ < $\frac{1}{2}$

www.mmhmath.com
For Real World Math Activities

Home Activity

Cut two shapes out of paper, such as a rectangle and a square.

Ask your child to fold each shape to make two or more equal parts.

Then have your child tell how many equal parts each shape has.

Books to Read

Look for these books at your local library and use them to help your child learn fractions.

- **Clean–Sweep Campers** by Lucille Recht Penner, The Kane Press, 2000.
- **Two Greedy Bears** by Mirra Ginsburg, Simon & Schuster, 1998.
- **Jump, Kangaroo, Jump!** by Stuart J. Murphy, HarperCollins, 1999.

Name_____

Unit Fractions

Learn A fraction can name a part of a whole.
Use a fraction to name equal parts.

Math Words
fraction
equal parts

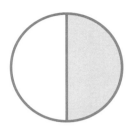

1 of 2 equal parts is pink.
One half is pink.

$\frac{1}{2}$ $\frac{1 \text{ pink part}}{2 \text{ equal parts}}$

1 of 3 equal parts is pink.
One third is pink.

$\frac{1}{3}$ $\frac{1 \text{ pink part}}{3 \text{ equal parts}}$

1 of 4 equal parts is pink.
One fourth is pink.

$\frac{1}{4}$ $\frac{1 \text{ pink part}}{4 \text{ equal parts}}$

1 of 8 equal parts is pink.
One eighth is pink.

$\frac{1}{8}$ $\frac{1 \text{ pink part}}{8 \text{ equal parts}}$

Try It Write the fraction.

1.  $\frac{1}{4}$ number of shaded parts / number of equal parts

2. 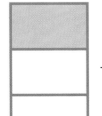   number of shaded parts / number of equal parts

3. **Write About It!** How can you tell if the parts are equal?

Chapter 25 Lesson 1   four hundred seventy-three **473**

Practice Write the fraction for the shaded part.

Write the number of shaded parts on top. Write the number of equal parts on the bottom.

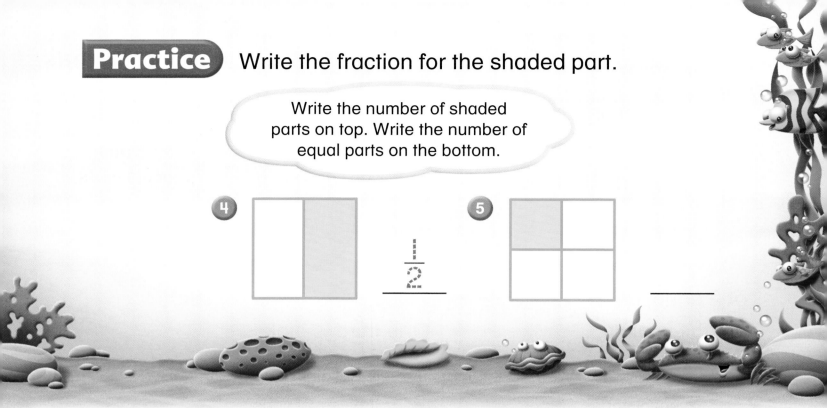

Color part of each shape to show the fraction.

6 $\frac{1}{3}$

7 $\frac{1}{8}$

8 $\frac{1}{4}$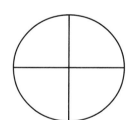

9 $\frac{1}{2}$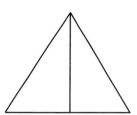

Make it Right

10 Dave says $\frac{1}{2}$ of the square is red.

Why is Dave wrong? Make it right.

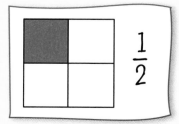

Math at Home: Your child learned about fractions as parts of a whole.
Activity: With your child, fold a sheet of paper into four equal parts. Have your child shade one part and then write a fraction to tell about his or her picture.

474 four hundred seventy-four

Name_____

Fractions Equal to 1

Learn You can write a fraction for the whole.

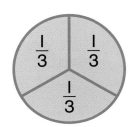

There are 3 green parts.
There are 3 equal parts.

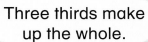

$\frac{3}{3}$ $\frac{\text{3 green parts}}{\text{3 equal parts}}$

The fraction for the whole is $\frac{3}{3}$.

The fraction for the whole always equals 1.

$\frac{3}{3} = 1$

Try It Count the parts in each whole.
Then write the fraction for the whole.

1.  $\frac{2}{2}$

2. 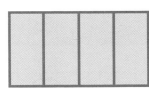 _____

3.  _____

4.  _____

5. **Write About It!** Why does a fraction for a whole have the same number on the top and the bottom?

Chapter 25 Lesson 2 · four hundred seventy-five **475**

Practice Count the parts in each whole.
Color the parts 🖍.
Then write the fraction for the whole.

The same whole can be made up of different numbers of parts.

6. $\frac{8}{8}$ is the same as 1.

$\frac{8}{8}$

7. ___

8. ___

9. ___

10. ___

11. ___

12. ___

Problem Solving — Reasoning

13. Each pizza shows halves.
Show how you count the halves.

$\frac{2}{2}$ $\frac{4}{2}$ $\frac{}{2}$ $\frac{}{2}$

Math at Home: Your child learned about fractions for the whole.
Activity: After you cut a sandwich in equal halves or fourths for your child, ask your child to name the fraction for the whole. (2 or 4)

Name _____

Problem Solving Practice

Solve.

1. There are 3 crabs. One is blue. What fraction is blue? Circle.

   $\frac{1}{2}$ $\frac{1}{3}$ $\frac{2}{3}$

2. 4 🐟 swim in the sea.
   4 🐟 swim away.

   Circle the fraction that shows what part of the 🐟 swim away.

   $\frac{1}{3}$ $\frac{5}{8}$ $\frac{4}{4}$

3. 6 divers look at fish. 4 of the divers are men. Write the fraction for the part of the divers that are men.

   _____

   What fraction of the divers are women?

   _____

4. Jen and Dad caught 12 fish all together. Jen caught 5 of them. Write the fraction for the fish Jen caught.

   _____

   What fraction of the fish did Dad catch?

   _____

 Write a Story!

5. Write a problem about a group of 8 sea turtles. Use the fraction $\frac{5}{8}$.

   _____

   _____

Chapter 25 Problem Solving Practice

Writing for Math

Jason, Tony, and Mary shared a pizza. Jason ate $\frac{1}{4}$ of the pizza. Tony ate $\frac{1}{2}$. Mary ate $\frac{1}{8}$. Who ate the most?

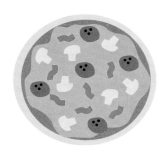

Think

The greatest fraction names the largest part.
I need to find the greatest fraction.
I can fold a paper circle to show fractions.

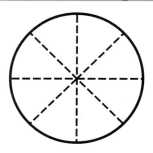

Solve

Color the paper circle to show each fraction.
Then I write the fractions in order from greatest to least.

_____

Explain

I can tell you who ate the most.

_____
_____
_____
_____

UNIT 7
CHAPTER 26

Probability

Heads or Tails?

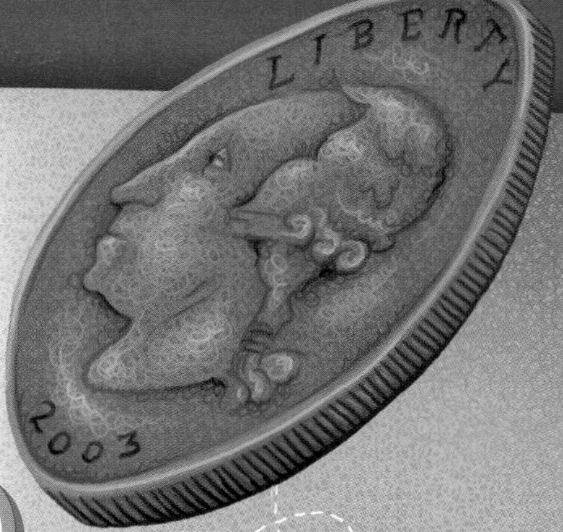

I toss a quarter in the air

To see which side will show.

Sometimes heads, and sometimes tails —

I never seem to know!

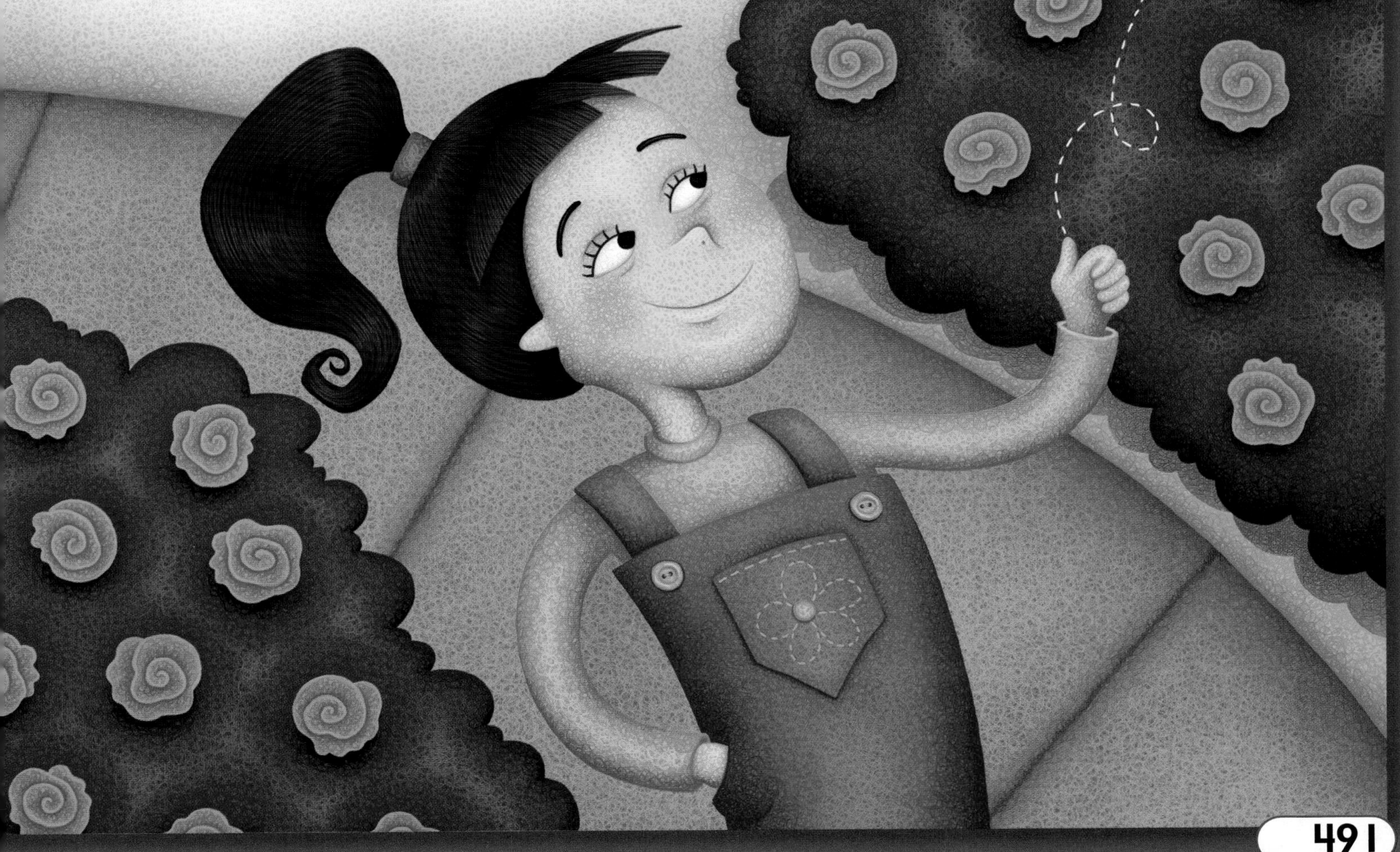

Math at Home

Dear Family,

I will learn about probability in Chapter 26. Here are my math words and an activity that we can do together.

Love, _____

My Math Words

equally likely :
  same chance to happen

less likely :
  not as probable

more likely :
  probable

prediction :
  telling before an event happens

I will spin red again.

www.mmhmath.com
For Real World Math Activities

Home Activity

Put 12 pennies on the table.

Have your child separate them into 2 equal groups.

Ask how many pennies there are in each group and what fraction each group represents.

Repeat the activity with 3, 4, and 6 equal groups.

Books to Read

Look for these books at your local library and use them to help your child learn probability.

- **Bad Luck Brad** by Gail Herman, The Kane Press, 2002.
- **No Fair!** by Caren Holtzman, Scholastic, 1997.
- **If You Give a Pig a Pancake** by Laura Numeroff, HarperCollins, 2000.

Name_____

Explore Probability

Learn You can tell if an event is certain, probable, or impossible.

Math Words
certain
probable
impossible

Certain	Probable	Impossible
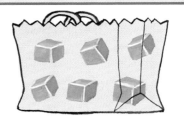		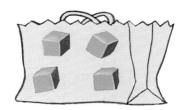
Picking an orange cube from this bag is certain. It will always happen.	Picking an orange cube from this bag is probable. It is likely to happen.	Picking an orange cube from this bag is impossible. It will never happen.

Try It Circle the answer.

1. Picking an orange cube is

   (certain)

   probable

   impossible

2. Picking a yellow cube is

   certain

   probable

   impossible

3. Picking a yellow cube is

   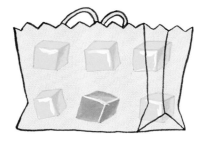

   certain

   probable

   impossible

4. Picking an orange cube is

   certain

   probable

   impossible

5. **Write About It!** Tell the colors of 8 cubes if picking orange is certain.

Practice Look at each spinner to answer the question. Circle the answer.

6 The spinner landing on blue is

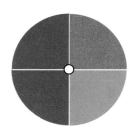

certain

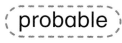

impossible

7 The spinner landing on red is

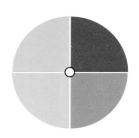

certain

probable

impossible

8 The spinner landing on yellow is

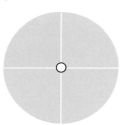

certain

probable

impossible

9 The spinner landing on green is

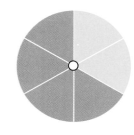

certain

probable

impossible

10 The spinner landing on blue is

certain

probable

impossible

11 The spinner landing on green is

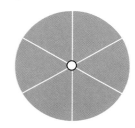

certain

probable

impossible

Problem Solving — Visual Thinking

Use the picture to answer the question.

12 Which color is impossible to pick? Explain your answer.

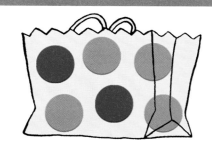

_____

Math at Home: Your child discussed the likelihood of a given event using the words *certain*, *probable*, and *impossible*.
Activity: Take a plastic bag. Put 5 pennies inside. Ask your child to use the word *certain*, *probable*, or *impossible* that he or she would pick a penny from the bag. (certain) A dime? (impossible)

494 four hundred ninety-four

Name_____

More Likely, Equally Likely, or Less Likely

Learn Sometimes an event is more likely, equally likely, or less likely to happen.

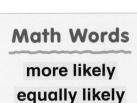

Math Words
more likely
equally likely
less likely

Toby is about to pick a triangle without looking.

Which color triangle is Toby more likely to pick?

Which is he less likely to pick?

What if there are the same number of and ?

Toby is ___equally likely___ to pick ▲ or ▲.

Your Turn

1. Look at each picture.

    Put cubes in a bag.

    • Which color are you more likely to pick? Color.

    • Which color are you less likely to pick? Color.

    • Then pick one cube without looking. Color the cube.

Bag	More Likely	Less Likely	Your Pick

2. **Write About It!** If there were 4 and 2 , would Toby more likely pick a ▲ or ▲? Why?

Chapter 26 Lesson 2

Practice

3 Look at each picture.

Put cubes in a bag.

- Which color are you more likely to pick? Color.
- Which color are you less likely to pick? Color.
- Then pick one cube without looking. Color the cube.

Bag	More Likely	Less Likely	Your Pick

Problem Solving — Visual Thinking

4 Color the spinner using red and blue so that red or blue are equally likely to be spun. Tell how you decide.

_____

_____

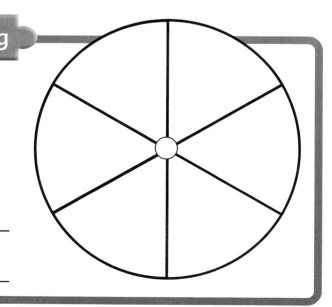

Math at Home: Your child learned about more likely, equally likely, and less likely outcomes.
Activity: Have your child put 4 dimes and 2 pennies in a paper bag. Ask your child to tell you which coin he or she is more likely to pick from the bag without looking and explain why. Then have your child do the activity.

Name_____   **Make Predictions**

Learn You can make a prediction that something will happen.

"I predict you will spin yellow again."

Math Word
prediction

Your Turn Use red and yellow to make your own spinner. Then answer the questions.

1. What color do you predict you will spin more often? _____

2. Predict. If you spin the spinner 12 times, how many times will you get red? _____

Spin the spinner 12 times. Record each spin.

red	yellow

3. How many times did the spinner land on yellow? _____

4. How many times did the spinner land on red? _____

5. **Write About It!** Will your predictions always match what happens?

Chapter 26   Lesson 3   four hundred ninety-seven **497**

Practice Use a penny.
Answer the questions.

6. Predict. If you toss the penny 10 times, how many times will you get

 ? _____   ? _____

Toss the penny 10 times. Record each toss.

7. How many times did you land on ? _____   ? _____

	1	2	3	4	5	6	7	8	9	10

Spiral Review and Test Prep

Choose the best answer.

8. Which picture shows a pie divided in fourths?

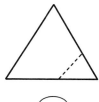   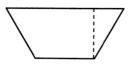

○   ○   ○   ○

9. Which picture shows a line of symmetry?

○   ○   ○   ○

Math at Home: Your child predicted outcomes.
Activity: Toss a nickel and have your child record how it landed. Repeat five times. Ask your child to predict how it will land next.

498 four hundred ninety-eight

Name _____

Extra Practice

Add or subtract.

42 +22	27 +25	49 +18	24 −15	73 −18
45 −21	62 +15	59 −19	32 +12	15 +19

Color in the quilt to complete the pattern.

1. ✏️ **Write About It!** Draw a pattern using a circle, square, and triangle. Repeat it.

LOG ON www.mmhmath.com
For more Practice

Math at Home: Your child practiced finding and extending patterns.
Activity: Draw a simple green and blue repeating pattern. Ask your child to draw what would most likely come next.

500 five hundred

Name _____

Problem Solving Strategy

Make a List

Sometimes you need to organize information.

Tracey is making a game with 2-digit numbers. She uses the digits 3, 4, and 6 to make different 2-digit numbers. How many different numbers can she make?

Read

What do I already know? _____

What do I need to find out? _____

Plan

I need to find how many different numbers Tracey can make. I will make a list.

Use each digit only once in the 2-digit number.

Solve

I can carry out my plan.

The list shows how many different 2-digit numbers.

There are __6__ numbers.

Digits	2-Digit Numbers
3	34 36
4	43 46
6	63 64

Look Back

Did I list all the numbers? _____

Chapter 26 Lesson 4

five hundred one **501**

Read | Plan | Solve | Look Back

Make a list to solve.

1. Emily uses the digits 2, 5, and 7 to make different 2-digit numbers. What numbers can she make?

Digits	2-Digit Numbers
2	___ ___
5	___ ___
7	___ ___

2. Trebor uses the digits 2, 4, and 8 to make different 2-digit numbers. What numbers can he make?

Digits	2-Digit Numbers
2	___ ___
4	___ ___
8	___ ___

3. Tracey's brother made different 2-digit numbers with 1, 5, and 9. What numbers can he make?

Digits	2-Digit Numbers
1	___ ___
5	___ ___
9	___ ___

4. Jason made different 2-digit numbers with 1, 2, 3, and 4. What numbers can he make?

Digits	2-Digit Numbers
1	___ ___ ___
2	___ ___ ___
3	___ ___ ___
4	___ ___ ___

Math at Home: Your child learned to solve problems by making an organized list.
Activity: Have your child make a list to show all the different 2-digit numbers he or she can make using 3, 7, and 9.

Name_____

Game Zone

Practice at School ★ Practice at Home

Name a Fraction!

▶ Take turns. Put the cubes in the bag. Shake the bag and then take out 4 cubes.
▶ Put a ● over the fraction that names the red part.
▶ Put the cubes back in the bag.
▶ Play until both players complete the chart.

2 players

You Will Need

3 ▫
5 ▪
8 ●

a paper bag

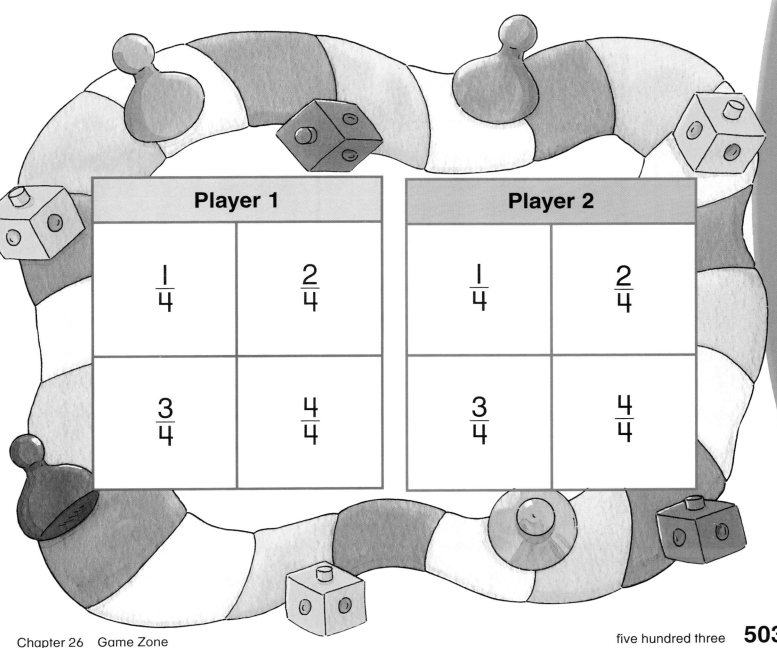

Player 1	
$\frac{1}{4}$	$\frac{2}{4}$
$\frac{3}{4}$	$\frac{4}{4}$

Player 2	
$\frac{1}{4}$	$\frac{2}{4}$
$\frac{3}{4}$	$\frac{4}{4}$

Technology Link

Compare Data • Calculator

You can use a to compare data.

Tim plays 3 games with Jake.

You Will Use

	Game 1	Game 2	Game 3
Tim	23	45	17
Jake	38	31	14

Who scored more points? How many more?

First you add Tim's score. Press.

   85

Record it. Then press .

Next you add Jake's score. Press.

[3] [8] [+] [3] [1] [+] [1] [4] [=]   83

__Tim__ scored __2__ more points than __Jake__.

85 > 83;
85 − 83 = 2

Use data from the chart to solve.

You can use a .

	Game 1	Game 2	Game 3	Game 4
Amy	32	16	17	25
Kim	29	13	24	27

1 Who scored more points? How many more?

_____ scored _____ more points than _____.

Chapter 26
Review/Test

Name _____

Circle the answer.

1) If you had a bag with only 2 blue marbles, picking a red marble from the bag is

certain   probable   impossible

2) If you had a bag with 2 black marbles and 9 pink marbles, picking a pink marble from the bag is

certain   probable   impossible

Answer each question.

3) Kwan has a bag of toy fish. There are 8 red and 4 yellow fish. Without looking, is he more likely or less likely to pick a yellow fish? _____

4) What color do you predict you will spin more often? Why?

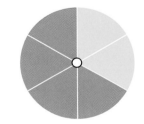

_____

_____

Solve.

5) Ashley uses the numbers 3, 7, and 9 to make different 2-digit numbers. What numbers can she make?

_____

Spiral Review and Test Prep
Chapters 1–26

Choose the best answer.

1 About how much does the glass hold?

- I fluid ounce
- 8 fluid ounces
- I quart
- I gallon

2 What time is it?

- 11:15
- 11:45
- 12:15
- 12:45

3 How much money?

- 46¢
- 55¢
- 56¢
- 66¢

Solve.

4 30 people were on the bus. Some people got off at Smith Street. Now 14 people are on the bus. How many people got off at Smith Street?

_____ people

5 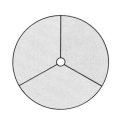 Which color will the spinner more likely land on? Explain how you know.

_____

www.mmhmath.com
For more Review and Test Prep

506 five hundred six

Interpreting Data

 ## The Giraffe Graph
by Sandra Liatsos

"My son," said the mother giraffe,

"very soon you'll grow bigger by half.

Each month we will measure

your height. What a pleasure

to show each new inch on a graph."

"I'll draw myself," said the giraffe,

"growing taller each month on my graph.

I'll soon be so tall

I'll go right off the wall,

and that will make both of us laugh."

Math at Home

Dear Family,

I will read and interpret sets of data in Chapter 27. Here are my math words and an activity that we can do together.

Love, _____

My Math Words

range: the difference between the greatest and least numbers

5 8 4 8 7 8 9
9 − 4 = 5
range = 5

mode: the number that occurs most often

5 8 4 8 7 8 9
mode = 8

median: the middle number when numbers are put in order from least to greatest

4 5 7 8 8 8 9
median = 8

www.mmhmath.com
For Real World Math Activities

Home Activity

Have your child think of a question to ask five people, such as "What is your favorite pet?" Then have your child record the results of the survey and discuss the information with you.

Favorite Pets	Tally	Total
cat	I	1
dog	III	3
bird	I	1

Books to Read

Look for these books at your local library and use them to help your child learn about different ways to show data.

- **How Many Snails?** by Paul Giganti, Morrow, William, & Company, 1994.
- **X Marks the Spot** by Lucille Recht Penner, The Kane Press, 2002.
- **The Button Box** by Margarette S. Reid, Puffin Books, 1990.

Name_____

Range and Mode

Learn You can describe a set of numbers by finding the range and mode of the numbers.

Math Words
range
mode

Al, Ken, Dennis, May, and Jan counted the letters in their names.

The range is the difference between the greatest and least numbers.

Subtract to find the range.

$6 - 2 = 4$.

The range is __4__.

The mode is the number that you see most often.

You see 3 most often.

The mode is __3__.

Try It Use these names. Answer each question.

1. How many letters are in the longest name? __5__

2. What is the range of the numbers? _____

3. Which number appears most often? _____

4. What is the mode of the numbers? _____

5. ✏️ **Write About It!** How do you find the range for a set of numbers?

Chapter 27 Lesson 1

five hundred nine **509**

Practice Mr. Park's class took a survey of the number of teeth they lost.

Range = difference between the greatest and least numbers.
Mode = number you see most often.

How Many Teeth We Lost	
Name	Number
Mary	5
Gus	7
Sue	12
Mike	9
Linda	2
Joe	7

6 What is the range of the numbers? __10__

7 What is the mode of the numbers? _____

The nature museum sold this many tickets in the last 4 days.

17   42   22   22

8 What is the range of the numbers? _____

9 What is the mode of the numbers? _____

Problem Solving — Use Data

Solve. Explain your answer.

10 A store sold 3 pies on Monday, 13 pies on Tuesday, 15 pies on Wednesday, 3 pies on Thursday, and 20 pies on Friday. What is the range of the numbers?

11 The cards show the numbers of books checked out by each student. What is the mode of the numbers?

Amy	Tom	Joe	Kim	Lin
4	3	1	10	3

Math at Home: Your child practiced finding the range and mode of a set of data.
Activity: Ask your child to record the age for each family member. Ask him or her to tell you what the range is. Is there a mode? What is it?

Name_____   **Median**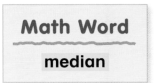

Learn You can use the median to describe a set of data.

What is the median age of the members?
You can use cubes to help you find the median for a set of data.

Ages of Club Members	
Name	Age (year)
Tom	8
Randy	7
Joe	9
Lisa	11
June	6

Math Word

median

Step 1
Make 5 cube towers.

8 7 9 11 6

Step 2
Put the towers in order from least to greatest number of cubes.

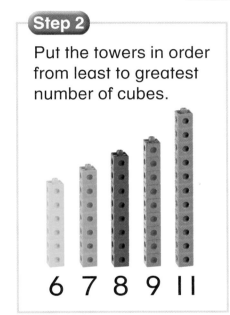

6 7 8 9 11

Step 3
Find the middle tower. The median is the number that is in the middle.

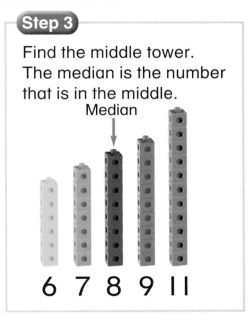

6 7 8 9 11

The median age of the club members is ___8___ years old.

Your Turn Find the median. Use cubes.

1. Put the towers in order from least to greatest number of cubes.

2. Find the median. _____

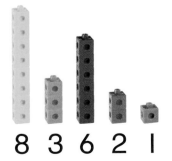

8 3 6 2 1

3. ✏️ **Write About It!** How do you find the median of a set of numbers?

Chapter 27 Lesson 2   five hundred eleven **511**

Practice Make towers with the following number of cubes. Find the median.

4) 4 cubes, 2 cubes, 9 cubes, 5 cubes, 7 cubes

Median: _____

The median is the number that is in the middle.

5) 3 cubes, 1 cube, 8 cubes, 10 cubes, 6 cubes

Median: _____

6) 4 cubes, 2 cubes, 7 cubes, 6 cubes, 3 cubes

Median: _____

7) 5 cubes, 2 cubes, 13 cubes, 7 cubes, 10 cubes

Median: _____

8) 6 cubes, 1 cube, 6 cubes, 8 cubes, 8 cubes

Median: _____

Problem Solving — Use Data

Solve. Explain your answer.

9) Josh found bugs every day for 5 days. Here are the numbers for each day.

8   6   10   12   9

What is the median of the numbers?

_____

10) Amy collected stickers for 5 days. Here are the numbers for each day.

20   2   15   11   1

What is the median of the numbers?

_____

Math at Home: Your child practiced finding the median of a set of numbers.
Activity: Have your child arrange 8 books in 3 stacks. Ask him or her to record the numbers in each stack and then tell you what the median is.

Name_____

Coordinate Graphs

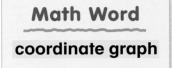

Learn A coordinate graph can show you where things are located.

This graph shows you where things are in a park. Find the sandbox.

- Always start at 0.
- First count to the right →.
- Then count up ↑.
- To find the sandbox go to the right 1 and up 2.

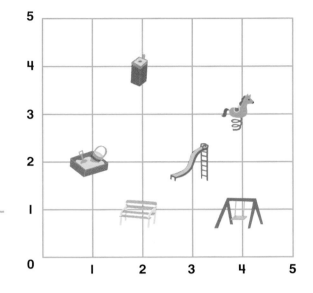

Try It Which thing would you find? Circle the answer.

	Right →	Up ↑			
1	2	4	sandbox	(fountain) ○	slide
2	4	3	horse	slide	swing
3	2	1	bench	sandbox	slide
4	3	2	fountain	slide	sandbox

5. ✏️ **Write About It!** Where is the swing set? Tell how you would get there.

Practice Where is each animal? Write the numbers.

		Right →	Up ↑
6	giraffe	3	3
7	zebra	___	___
8	bird	___	___
9	elephant	___	___
10	monkey	___	___

Problem Solving — Number Sense

11 An apple costs 75¢. Larry does not have enough money to buy it. He needs 10¢ more. How much money does Larry have? Explain.

_____

12 An orange costs 40¢. Sue needs 15¢ more. How much money does she have? Explain.

_____

Math at Home: Your child learned how to locate and write points on a grid.
Activity: Name an object on the grid on page 513. Have your child name the points where it is found using the words *go right* and *go up*.

Name_____

Line Graphs

Math Word

line graph

 You can use a line graph to compare data over time.

Adam wrote the daily high temperatures for four days.

Temperature	
Day	**°F**
Day 1	70°
Day 2	60°
Day 3	70°
Day 4	80°

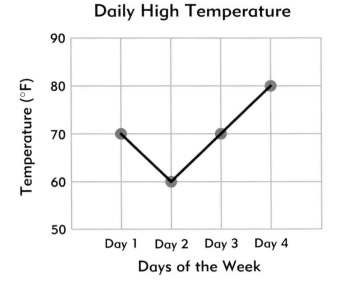

Then Adam drew a dot for each day.
He connected the dots to make a line graph.

Try It Complete the line graph.

1. Anna measured her puppy every month for 4 months.

Puppy Height	
Month	**Height in Inches**
March	7
April	8
May	8
June	9

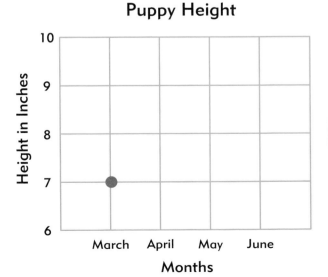

2. **Write About It!** Is the puppy bigger in June than in March? How do you know?

Chapter 27   Lesson 4   five hundred fifteen **515**

Practice Make a line graph to show the data.

Draw a dot for each day.

Sam wrote the temperature at noon for four days.

Daily Noon Temperature	
Day	Temperature
Thursday	50°
Friday	55°
Saturday	60°
Sunday	50°

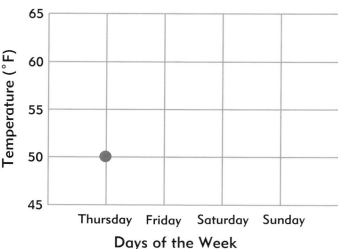

Use the line graph above to solve.

3 Was it warmer at noon on Thursday or Friday?

_____

4 On which day was it the warmest at noon?

_____

Problem Solving — Use Data

Use the Venn diagram to answer the questions.

5 How many children have a pet? _____

6 How many children have a sister and a pet? _____

7 How many children have a brother, sister, and a pet? _____

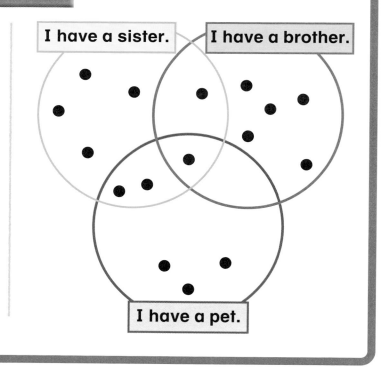

Math at Home: Your child learned how to make and interpret line graphs.
Activity: Ask your child to explain how he or she made the graph on this page.

Problem Solving Skill
Reading for Math

Boxes and Boxes

There are 5 trucks in the lot behind the supermarket. They bring fruits and vegetables from farms outside the city. The graph below shows what they brought.

Fruits and Vegetables

Beans								
Carrots								
Corn								
Apples								
Bananas								
Oranges								
	0	1	2	3	4	5	6	7	8

Number of Boxes

Reading Skill: Important and Unimportant Information

1. Was the number of trucks important to the story? _____

2. How many boxes of carrots were there? _____

3. How many boxes of bananas were there? _____

4. Are there more boxes of corn or apples? _____

Chapter 27 Lesson 5

More Boxes

Even though it's raining, more trucks arrived. The graph shows what they brought. Next week, the trucks will return to deliver more boxes.

Things to Buy

	0	1	2	3	4	5	6	7	8
Bread									
Cereal									
Paper Towels									
Popcorn									
Soap									
Tacos									

Number of Boxes

Important and Unimportant Information

5 What information in the story was unimportant to help you read the graph?

_____

6 How many boxes of paper towels were there? _____

7 The trucks delivered 6 boxes of two items. What were they?

_____

Math at Home: Your child identified information that was not needed to answer questions.
Activity: Have your child cross out the unimportant sentences in each story.

Name _____

Problem Solving Practice

Solve.

1) How many children like to 🚲?

_____ like to 🚲

Favorite Things to Do

(bar graph showing: bicycle = 5, computer = 4, book = 3, skateboard = 1)

0 1 2 3 4 5
Number of Votes

2) Do more children like The Frog Prince or Puss in Boots?

_____

How many more?

_____

Favorite Fairy Tales

Rapunzel							
Cinderella							
Puss in Boots							
The Frog Prince							

0 1 2 3 4 5 6 7
Number of Votes

Write a Story!

3) Write a question about this data. Answer your question.

_____

_____

Favorite Pets

Dog						
Bird						
Cat						
Fish						

0 1 2 3 4 5 6
Number of Votes

Chapter 27 Problem Solving Practice

Writing for Math

Rosa puts these dots on a coordinate graph.

Right →	Up ↑
1	2
2	4
3	2

If she connects them what shape will she make?

Think

Always start at 0.

First count to the right →.

Then count up ↑.

Solve

Put the dots on the coordinate graph. Then connect the dots.

Rosa will make a _____.

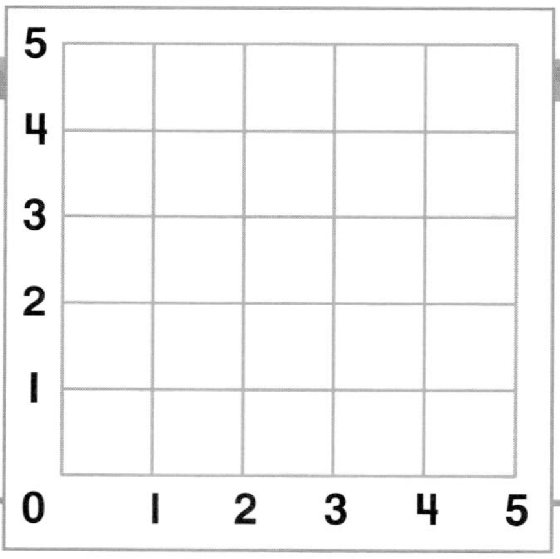

Explain

I can tell you how I found the points and drew the shape.

_____

_____

Chapter 27
Review/Test

Name_____

Use the data to answer each question.

Dance Tickets Sold	
Monday	25
Tuesday	18
Wednesday	25
Thursday	19
Friday	14

1. What is the range of the numbers?

   _____

2. What is the mode of the numbers?

   _____

3. The dance was 2 hours long on Friday. How many tickets in all were sold on Friday and Monday?

   _____

Find each object. Complete to show where it is.

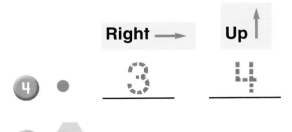

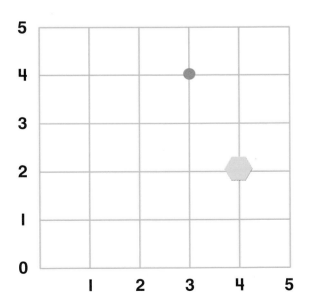

Spiral Review and Test Prep
Chapters 1–27

Use this graph to answer problems 1–2.

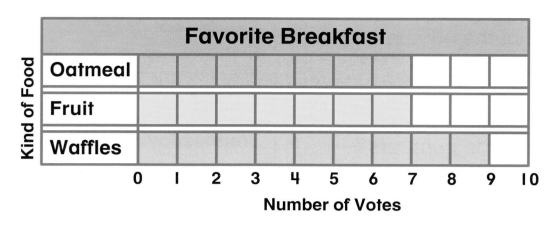

1. How many like oatmeal?

   7   9   10   11
   ○   ○   ○   ○

2. How many more like waffles than fruit?

   1   2   7   9
   ○   ○   ○   ○

3. What is the range of these numbers?

   2, 7, 6, 12, 5 _____

4. Write 3 facts about the graph.

   _____

   _____

   _____

Favorite Color	
Green	★
Yellow	★★★★
Purple	★★
Red	★★★
Each ☆ stands for 2 votes.	

www.mmhmath.com
For more Review and Test Prep

UNIT 7 CHAPTER 28

Exploring Multiplication and Division

Caterpillar Pete

by Sandra Liatsos

"If I could multiply times two,"
said Caterpillar Pete,
"I'd figure out
how many shoes
I needed for my feet.
On winter mornings every foot
could snuggle in its slipper.
On summer mornings in the sea
each foot could flip a flipper."

"If I could multiply times two,"
the caterpillar cried,
"I'd know how many shoes to buy
by counting just one side!"

Math at Home

Dear Family,

I will multiply and divide in Chapter 28. Here are my math words and an activity that we can do together.

Love, _____

My Math Words

multiply:

$2 \times 4 = 8$

divide:

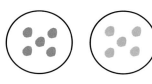

$10 \div 2 = 5$

Home Activity

Make a set of cards with addition sentences on them, such as $5 + 5 = 10$, and $2 + 2 + 2 + 2 = 8$.

Arrange 10 straws on the table in 2 groups of 5.

Ask your child to find the addition sentence that shows how many straws in all. Repeat the activity with other equal groups of straws.

Books to Read

In addition to these library books, look for the Time For Kids math story that your child will bring home at the end of this unit.

- **My Full Moon Is Square** by Elinor Pinczes, Houghton Mifflin, 2002.
- **The Doorbell Rang** by Pat Hutchins, Morrow, William & Company, 1992.
- **Time For Kids**

www.mmhmath.com
For Real World Math Activities

524 five hundred twenty-four

Name_____

Explore Equal Groups

Learn When there are **equal groups**, you can **skip-count** to find the total.

Math Words
skip-count
equal group

__2__, __4__, __6__, __8__ counters in all

You can make equal groups.
Put two counters in each group.
How many equal groups?

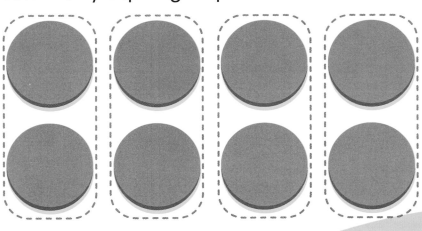

__4__ equal groups

Your Turn Use ●. Skip-count to find the total.

1. Make 2 groups of 4.

   __8__ ● in all

2. Make 4 groups of 3.

   _____ ● in all

3. **Write About It!** When can you skip-count to find how many in all?

Chapter 28 Lesson 1 — five hundred twenty-five **525**

Practice Skip-count. Write how many in all.

The bugs are in equal groups.

       in all

_____ _____ _____ _____ _____ in all

_____ _____ _____ _____ _____ in all

How many equal groups can you make?
Circle the groups. Write how many.

_____ equal groups

_____ equal groups

Math at Home: Your child learned to skip-count equal groups.
Activity: Give your child a pile of nickels and have him or her skip-count by 5s. Then ask the child to put the coins in piles to make equal groups.

Name _____

Repeated Addition and Multiplication

Learn When groups are equal, you can use repeated addition to find how many in all. Or you can multiply.

Math Word
multiply

I can write an addition sentence.

__2__ + __2__ + __2__ + __2__ + __2__ = __10__

I can also write a multiplication sentence.

__5__ × __2__ = __10__

Five groups of 2 are 10.

Your Turn Use 🔲 to make equal groups. Add. Then multiply.

1. Make 2 groups of 4.

   __4__ + __4__ = __8__

   __2__ × __4__ = __8__

2. Make 3 groups of 2.

   ____ + ____ + ____ = ____

   ____ × ____ = ____

3. Make 4 groups of 5.

   ____ + ____ + ____ + ____ = ____

   ____ × ____ = ____

4. ✏️ **Write About It!** How would you write a multiplication sentence for 3 + 3 + 3 + 3? Explain.

Practice Add. Then multiply.

Repeated addition and multiplication are related.

5.

__5__ + __5__ = __10__

__2__ × __5__ = __10__

6.

____ + ____ + ____ + ____ + ____ = ____

____ × ____ = ____

7.

____ + ____ + ____ + ____ + ____ + ____ = ____

____ × ____ = ____

Problem Solving **Number Sense**

8. Use the addition sentence to complete the multiplication sentence. Tell how you found your answer.

$4 + 4 + 4 + 4 + 4 = 20$

☐ × 4 = 20

_____

_____

Math at Home: Your child learned how repeated addition and multiplication are related.
Activity: Show 4 groups of three coins. Have your child write a multiplication sentence to find how many coins in all. (4 × 3 = 12)

Name_____

Use Arrays to Multiply

Learn You can use an array to help you multiply.

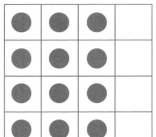

There are 4 rows in this array.
3 counters are in each row.
There are 12 counters in all.
$3 + 3 + 3 + 3 = 12$
or $4 \times 3 = 12$

Math Word
array

__4__ × __3__ = __12__
rows in each row in all

Your Turn Use counters and the grid to show the array for each multiplication sentence.

1. $2 \times 3 = \underline{6}$
   rows in each row in all

2. $4 \times 4 = \underline{}$
   rows in each row in all

3. $4 \times 2 = \underline{}$
   rows in each row in all

4. $2 \times 4 = \underline{}$
   rows in each row in all

5. **Write About It!** Look at exercises 3 and 4. Are the answers the same? Why or why not?

Chapter 28 Lesson 3

Practice Write a multiplication sentence for each array.

Use an array to help you multiply.

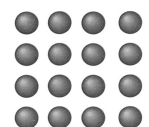

6) __4__ × __4__ = __16__
   rows   in each row   in all

7) _____ × _____ = _____
   rows   in each row   in all

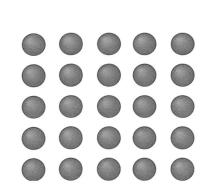

8) _____ × _____ = _____
   rows   in each row   in all

9) _____ × _____ = _____
   rows   in each row   in all

Problem Solving — Number Sense

Show Your Work

10) There are 2 rows of flowers. Each row has 4 flowers. How many flowers in all?

_____ × _____ = _____
rows   in each row   in all

Draw a picture to solve.

Math at Home: Your child practiced writing multiplication sentences using multiplication arrays.
Activity: Draw 2 rows of 3 dots. Have your child write a multiplication sentence to show the array. (2 × 3 = 6)

Extra Practice

Name_____

Add or multiply.

$2 + 2 + 2 = \boxed{6}$

$3 \times 4 = \boxed{}$

$2 \times 5 = \boxed{}$

$8 \times 2 = \boxed{}$

$6 \times 4 = \boxed{}$

$3 \times 6 = \boxed{}$

$6 \times 2 = \boxed{}$

$5 + 5 + 5 = \boxed{}$

$2 + 2 + 2 + 2 = \boxed{}$

$5 \times 3 = \boxed{}$

$2 \times 4 = \boxed{}$

$5 \times 5 = \boxed{}$

$2 \times 2 = \boxed{}$

$1 + 1 + 1 + 1 = \boxed{}$

$4 \times 1 = \boxed{}$

Chapter 28 Extra Practice

Circle the unit you would use to measure.

1. the amount of juice in a

    (cups)   gallons

2. the length of a

    inches   feet

3. how heavy a is

    grams   kilograms

4. the amount of milk in a

    ounces   gallons

5. the length of a

    inches   yards

6. the weight of a

    ounces   pounds

7. the amount of water in a

    ounces   gallons

8. the length of a

    inches   feet

9. the length of a

    centimeters   meters

10. the height of a

    inches   yards

11. Would you use inches or yards to measure the length of the playground? Explain why.

_____

 www.mmhmath.com
For more Practice

Math at Home: Your child practiced choosing the right unit to measure an object.
Activity: Choose an object in the kitchen. Ask your child to identify units for measuring its length and weight.

Name_____

Repeated Subtraction and Division

HANDS ON Activity

Learn You can use repeated subtraction to find the number of equal groups. Or you can divide.

Math Word
_____
divide

Use 🟦. Separate 12 cubes into equal groups of 4 each. How many equal groups of 4 can you make?

Subtract 4 three times. Make 3 groups of 4. Write a division sentence.

$12 - 4 = 8 \quad 8 - 4 = 4 \quad 4 - 4 = 0$

You get __3__ groups of 4.

__12__ ÷ __4__ = __3__

Your Turn Use 🟦. How many equal groups can you make? Subtract. Then divide.

① Use 10 🟦.
Subtract groups of 2.

You get __5__ groups of 2.

__10__ ÷ __2__ = __5__

② Use 6 🟦.
Subtract groups of 3.

You get ____ groups of 3.

____ ÷ ____ = ____

③ ✏️ **Write About It!** How does subtracting equal groups help you divide?

Practice Divide. You can draw a picture to help.

④ 12 bugs
3 in each jar

How many jars do you need?

$12 \div 3 = \underline{4}$

$\underline{4}$ jars

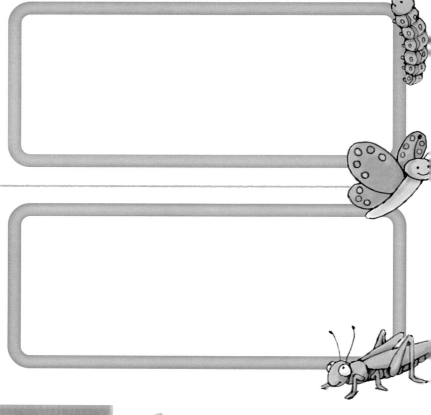

⑤ 16 worms
4 in each box

$16 \div 4 = \underline{}$

_____ boxes

⑥ 20 butterflies
4 in each net

$20 \div 4 = \underline{}$

_____ nets

Problem Solving — Critical Thinking

Show Your Work

⑦ Brandon catches 12 bugs and puts them in jars. He puts 2 bugs in each jar. How many jars does Brandon use?

_____ jars

Draw a picture to solve.

Math at Home: Your child practiced doing repeated subtraction to divide into equal groups.
Activity: Put 20 raisins or crackers on the table. Ask your child to divide them equally into groups of 5.

Name _____

Divide to Find Equal Shares

Learn You can divide to find equal shares.

Let's divide the crayons equally between us.

Math Word
equal share

$6 \div 2 = 3$

We each have __3__ crayons.

We each have an equal share.

Your Turn Use ●. Make equal groups.
Write how many in each group. Divide.

1. 4 ●

   2 equal groups

   __2__ in each group

   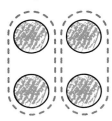

   $4 \div 2 = \underline{\ 2\ }$

2. 10 ●

   2 equal groups

   _____ in each group

   $10 \div 2 = \underline{\ \ \ \ }$

3. 12 ●

   4 equal groups

   _____ in each group

   $12 \div 4 = \underline{\ \ \ \ }$

4. **Write About It!** If you have 7 toys, can you make 2 equal groups? Explain.

Chapter 28 Lesson 5   five hundred thirty-five **535**

Practice Divide. You can draw a picture to help.

You can divide to find equal shares.

5) 12 flowers
6 equal groups
How many flowers in each group?

$12 \div 6 = \underline{2}$

6) 16 bugs
4 equal groups
How many bugs in each group?

$16 \div 4 = \underline{}$

7) 20 ants
4 equal groups of ants
How many ants in each group?

$20 \div 4 = \underline{}$

8) 15 butterflies
3 equal groups
How many butterflies in each group?

$15 \div 3 = \underline{}$

Spiral Review and Test Prep

9) Which number completes the multiplication?

$5 \times 8 = \boxed{}$

12   13   40   58

10) What is the mode?

4, 3, 2, 3, 1

2   3   4

Math at Home: Your child divided to find equal shares.
Activity: Show 10 buttons. Have your child make 5 equal groups and tell how many are in each group. (2 buttons)

Name_____

Problem Solving Strategy

Use a Pattern • Algebra

You can use a pattern to help you solve problems.

A duck has 2 feet. How many feet are on 6 ducks?

Read

What do I already know? _____ feet on 1 duck

What do I need to find out? _____

Plan

I can find a pattern.

Solve

I can carry out my plan.
I can make a chart.

Number of Ducks	1	2	3	4	5	6
Number of Feet	2	4				

There are _____ feet on 6 ducks.

Look Back

What pattern do I see? _____

Chapter 28 Lesson 6

five hundred thirty-seven **537**

Read Plan Solve Look Back

Use a number pattern to solve.

1. How many wings are on 5 butterflies?

   There are _____ wings on 5 butterflies.
   What pattern do I see?

Number of Butterflies	1	2	3	4	5
Number of Wings	4	8			

   _____

2. How many fingers are on 5 gloves?

   There are _____ fingers on 5 gloves.
   What pattern do I see?

Number of Gloves	1	2	3	4	5
Number of Fingers	5				

   _____

3. How many wheels are on 5 tricycles?

   There are _____ wheels on 5 tricycles.
   What pattern do I see?

Number of Tricycles	1	2	3	4	5
Number of Wheels	3				

   _____

4. How many legs are on 6 crabs?

   There are _____ legs on 6 crabs.
   What pattern do I see?

Number of Crabs	1	2	3	4	5	6
Number of Legs	10					

   _____

Math at Home: Your child solved problems by using number patterns.
Activity: Have your child continue one of the number patterns on this page.

Name _____

Game Zone

Practice at School ★ Practice at Home

Butterfly Bingo

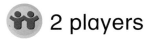 2 players

- You and a partner take turns. Choose blue or green for your counter.
- Toss both cubes.
- Multiply the two numbers and place your counter on the answer.

The first player to cover a row, column, or diagonal wins.

You Will Need

2 cubes
15 ●
15 ●

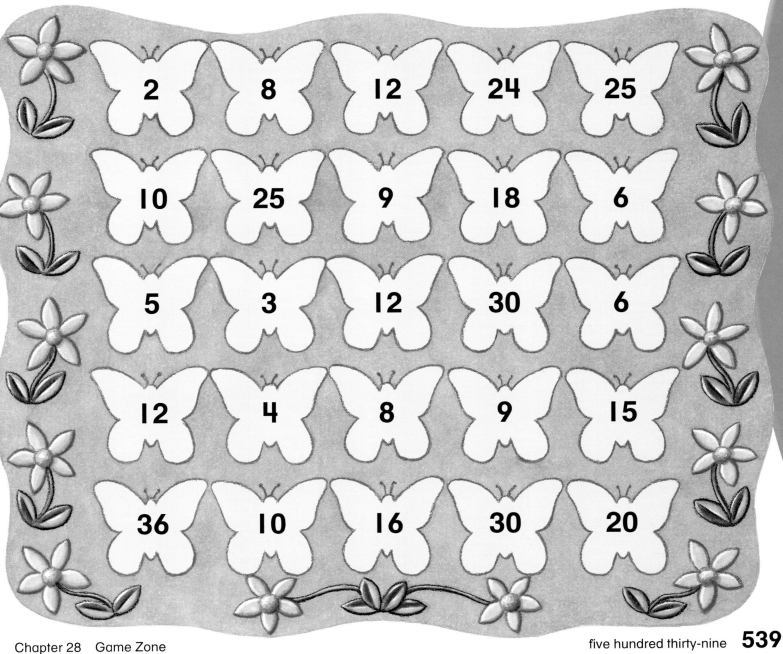

Technology Link

Model Multiplication • Computer

Use [icon].

- Choose a mat to show one number.
- Stamp out 2 rows of 5 butterflies.
- What multiplication fact do you see?

  __2__ × __5__ = __10__

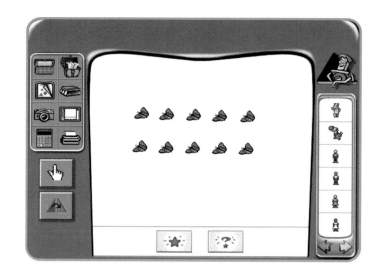

- Stamp out 3 rows of 4 butterflies.
- Write the multiplication fact you see.

  __3__ × __4__ = __12__

Use counters to make other multiplication facts.

____ × ____ = ____

____ × ____ = ____

____ × ____ = ____

____ × ____ = ____

____ × ____ = ____

____ × ____ = ____

 For more practice use Math Traveler.™

Chapter 28
Review/Test

Name_____

Write a multiplication sentence for each array.

1)

2)

____ × ____ = ____   ____ × ____ = ____

Write how many groups.

3) 16 beetles
4 on each leaf

____ groups

16 ÷ 4 = ____

Write how many in each group.

4) 8 butterflies
4 nets

____ in each group

8 ÷ 4 = ____

Use a pattern to solve.

5) There are 4 bugs. Each bug has 6 legs. How many legs in all?

____ legs

Number of Bugs	1	2	3	4
Number of Legs	6			

Spiral Review and Test Prep
Chapters 1–28

Choose the best answer.

1 What is the mode of these numbers?

3, 10, 15, 4, 10, 9, 12

- 9
- 10
- 12
- 15

2 How many feet are there in a yard?

- 1
- 3
- 10
- 12

3 What is the value of the 5 in 450?

- 5
- 50
- 55
- 500

4 Add. 519
        +275

5 Multiply.

$5 \times 2 =$ ____

6 If Mary takes a cube from the bag without looking, what cube is she more likely to pick?

_____ cube

Write to explain your answer.

Name _____

Two friends each eat one slice of the pie.
Write the fraction of the pie they eat.

_____

Color the pie to show the fraction.

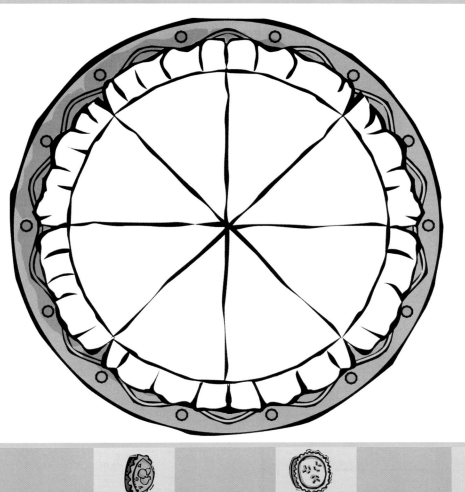

How Many Peaches?

Picking peaches with a friend is fun. You can pick peaches to make a pie.

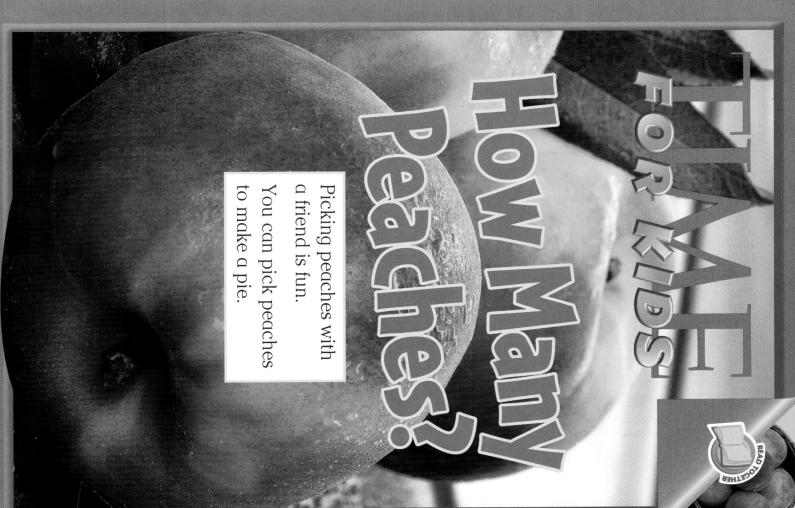

Mom's Peach Pie Recipe

What You Will Need

- A grown-up to help you
- 14 fresh peach halves or canned peaches
- $\frac{3}{4}$ cup sugar
- $\frac{1}{4}$ cup flour
- $\frac{1}{4}$ cup water or peach juice
- 2 tablespoons butter
- 2 tablespoons lemon juice
- 9-inch pie crust

What You Do

1. Mix sugar, butter, and flour to make crumbs.
2. Sprinkle half the mixture in the bottom of an unbaked pie crust.
3. Place peaches in the pie shell. Sprinkle with the lemon juice.
4. Cover with the rest of the crumbs.
5. Add fruit juice.
6. Bake at 375°F for 40–45 minutes.

Two friends picked 24 peaches in all. 6 peaches fit in a basket. They filled 4 baskets.

24 ÷ 6 = 4

Mom needs 12 peaches to make a pie. She will use 2 baskets of peaches.

2 × 6 = 12

Name _____

Problem Solving Decision Making

Use Fractions to Make Decisions

Plan a pizza party for 6 children.
Each child may eat 2 or more slices.
3 children only eat mushroom pizza.
Each pizza has 8 slices.

 ① Decide how many slices the children will eat. Circle to show.

 ② How many slices of each kind of pizza did you circle?

③ How many pizzas of each kind must you buy for the party? Color to show.

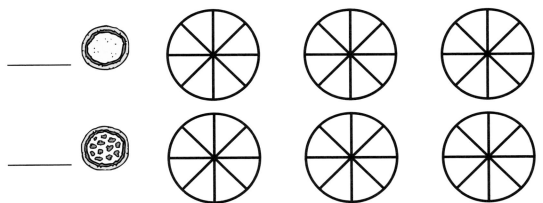

Unit 7 Decision Making

Plan a party for 12 children. Each child may have 2 or more party favors. There are 12 party favors to a box. Each box has one kind of party favor in it.

 4 Decide how many party favors you will give the children. Color to show.

 5 How many of each kind of party favor will you need?

6 How many boxes of each must you buy for the party? Color to show.

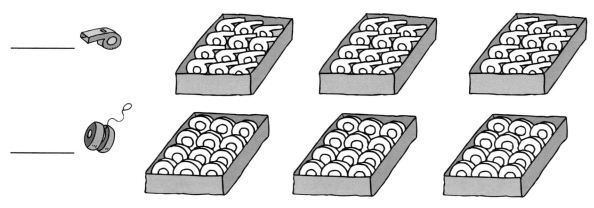

 **Math at Home:** Your child applied fraction concepts to make decisions.
Activity: Show your child a group of two kinds of coins, such as 3 pennies and 3 nickels. Ask your child to write the fraction for the number of pennies.

Name _____

Unit 7
Study Guide and Review

Math Words

Draw lines to match.

1. $4 \times 4 = 16$ — equal groups

2. 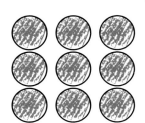 — multiplication sentence

3. $\frac{1}{3}$ — fraction

Skills and Applications

Fractions (pages 473-484)

Examples

Fractions show equal parts of a whole.

 $\frac{1}{4}$ 1 of 4 equal parts

 $\frac{2}{4}$ 2 of 4 equal parts

4. 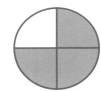 $\frac{\square}{4}$

5. $\frac{\square}{2}$

Fractions can show equal parts of a group.

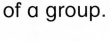

$\frac{3}{4}$ ← number of red fish
      ← total number of fish

6. $\frac{\square}{3}$

Unit 7 Study Guide and Review   five hundred forty-five **545**

Skills and Applications

Multiplication and Division (pages 525–536)

Examples

You can use addition to help you multiply.

$5 + 5 = 10$

$2 \times 5 = 10$

7. $4 \times 2 =$ _____

8. $6 \times 3 =$ _____

9. $5 \times 5 =$ _____

10. $4 \times 3 =$ _____

How many in each group?

10 counters

5 groups

$10 \div 5 = 2$ in each group

11. 15 counters

 3 in each group

 $15 \div 3 =$ _____ groups

12. 16 counters

 4 groups

 $16 \div 4 =$ _____ groups

Problem Solving — Strategy (pages 537–538)

13. Find the pattern.

Number of trucks	1	2	3	4	5
Number of wheels	4	8			

There are _____ wheels on 5 trucks.

What is the pattern?

_____

Math at Home: Your child learned about fractions, interpreting data, and multiplication and division.
Activity: Have your child use these pages to review fractions, interpreting data, and multiplication and division.

Name_____

Unit 7 Performance Assessment

Play a Game

Number of Bags   Number of Marbles

You Will Need

2 counters

Toss a cube on each mat to find the number of bags and marbles.

Use the numbers to find how many marbles in all.

Write a multiplication sentence. You can use counters if you need to.

1. ____ × ____ = ____ marbles in all

2. ____ × ____ = ____ marbles in all

3. ____ × ____ = ____ marbles in all

4. ____ × ____ = ____ marbles in all

 You may want to put this page in your portfolio.

Unit 7
Enrichment

Remainder

Sometimes when you divide, you have a remainder.

Show 17 tomatoes in 2 equal groups.

__8__ tomatoes in each group

__1__ tomato left   17 ÷ 2 = __8__ R __1__

Draw dots to show the equal groups and the remainder.

1 Show 13 tomatoes in 2 equal groups.

_____ tomatoes in each group

_____ tomato left   13 ÷ 2 = _____ R _____

2 Show 9 tomatoes in 4 equal groups.

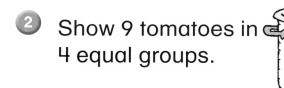

_____ tomatoes in each group

_____ tomato left   9 ÷ 4 = _____ R _____

3 Show 14 tomatoes in 3 equal groups.

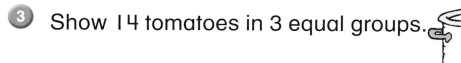

_____ tomatoes in each group

_____ tomatoes left   14 ÷ 3 = _____ R _____

Picture Glossary

add (+) (page 7)

2 + 3 = 5

$$\begin{array}{r}2\\+3\\\hline 5\end{array}$$

area (page 381)

The area of this rectangle is 6 square units.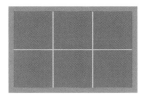

addend (page 17)

$$\begin{array}{r}31 \leftarrow \text{addend}\\+18 \leftarrow \text{addend}\\\hline 49\end{array}$$

array (page 529)

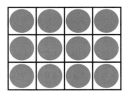

3 × 4 = 12

after (page 97)

47 48 49

48 is after 47.

bar graph (page 189)

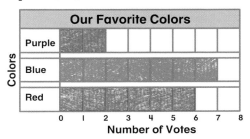

A.M. (page 171)

the hours from midnight to noon

It is 7:00 A.M.

before (page 97)

47 48 49

47 is before 48.

angle (page 355)

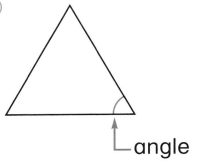

angle

between (page 97)

47 48 49

48 is between 47 and 49.

Glossary **G1**

Picture Glossary

calendar (page 177)

This is a calendar for September.

chart (page 191)

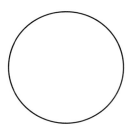

capacity (page 333)

the amount a container holds when filled

circle (page 355)

cent sign (¢) (page 115)

1¢   1 cent

compare (page 95)

5 is less than 7.   6 is equal to 6.   8 is greater than 4.

centimeter (cm) (page 323)

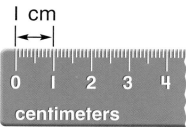

cone (page 353)

certain (page 493)

An event will definitely happen.

It is certain that you will pick a ▪.

congruent (page 373)

same size and same shape

Picture Glossary

coordinate graph (page 513)

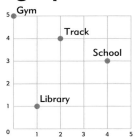

cup (c) (page 335)

count back (page 33)

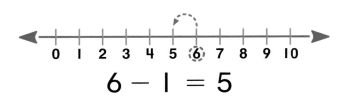

$6 - 1 = 5$

cylinder (page 353)

count on (page 19)

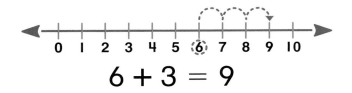

$6 + 3 = 9$

data (page 189)

information that is collected for a survey

counting pattern (page 5)

2, 4, 6, 8, 10, 12, 14	Counting by 2s
3, 6, 9, 12, 15, 18, 21	Counting by 3s
5, 10, 15, 20, 25, 30, 35	Counting by 5s

decimal point (page 133)

decimal point

cube (page 353)

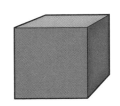

degrees Celsius (°C) (page 343)

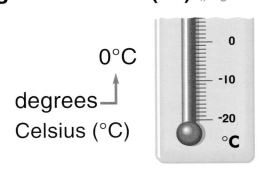

Glossary **G3**

Picture Glossary

degrees Fahrenheit (°F) (page 343)

16°F
degrees Fahrenheit (°F)

divide (page 533)

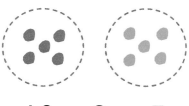

$10 \div 2 = 5$

difference (page 7)

$17 - 9 = 8$

$$\begin{array}{r} 17 \\ -9 \\ \hline 8 \end{array}$$

difference ⟶

Subtract to find the difference.

dollar ($) (page 133)

$1.00   100 cents

digit (page 81)

any single figure used when representing a number

354

3, 5, and 4 are digits in 354.

doubles (page 53)

$7 + 7 = 14$

dime (page 115)

10¢   10 cents

doubles plus 1 (page 53)

$7 + 8 = 15$

distance (page 323)

the space between two points

The distance between the 2 houses is 200 yards.

edge (page 353)

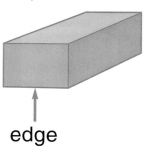

edge

Picture Glossary

equal groups (page 525)

There are 4 equal groups of counters.

equal parts (page 473)

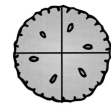

This pie is cut into 4 equal parts.

equal share (page 535)

 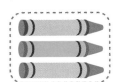

Two children can share these crayons equally.

equally likely (page 495)

Without looking, it is equally likely that you will pick a ▨ as a ▨.

estimate (page 259)

47 + 22

50 + 20

about 70 ←— estimate

even (page 105)

2 4 6 8 10

expanded form (page 401)

another way of writing a number

364

300 + 60 + 4 ←— expanded form

face (page 353)

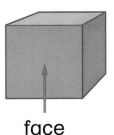

face

fact family (page 59)

$5 + 6 = 11 \quad 11 - 6 = 5$
$6 + 5 = 11 \quad 11 - 5 = 6$

flip (page 377)

a mirror image of a figure

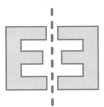

Picture Glossary

fluid ounce (fl oz) (page 335)

8 fluid ounces in a cup

foot (ft) (page 321)

12 inches = 1 foot

fraction (page 473)

$\frac{1}{2}$ $\frac{1}{3}$ $\frac{1}{4}$ $\frac{3}{4}$

gallon (gal) (page 335)

4 quarts = 1 gallon

half dollar (page 121)

50¢   50 cents

half hour (page 155)

30 minutes = half hour

hexagon (page 359)

6 sides and 6 angles

hour (page 155)

60 minutes = 1 hour

hour hand (page 155)

hundreds (page 397)

2 hundreds

G6

Picture Glossary

impossible (page 493)

an event that cannot happen

It is impossible to pick a ■.

inch (in.) (page 319)

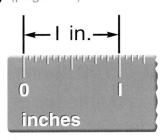

is equal to (=) (page 95)

35 is equal to 35.

35 = 35

is greater than (>) (page 95)

36 is greater than 35.

36 > 35

is less than (<) (page 95)

35 is less than 36.

35 < 36

key (page 195)

tells you what each symbol stands for

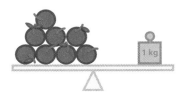

kilogram (kg) (page 341)

1 kilogram is about 8 apples.

length (page 319)

Length is how long something is.

less likely (page 495)

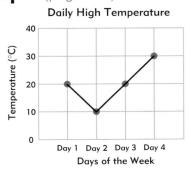

Without looking, it is less likely that you will pick a ■.

line graph (page 515)

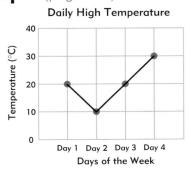

LOG ON www.mmhmath.com to find out about these words

Picture Glossary

line of symmetry (page 375)

a line on which a figure can be folded so that its two halves match exactly

median (page 511)

the middle number when numbers are put in order from least to greatest

6 7 8 9 10

The median is 8.

line plot (page 197)

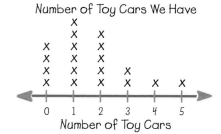

meter (m) (page 323)

1 meter = 100 centimeters

1 meter is a little longer than a baseball bat.

liter (L) (page 339)

There are 1,000 milliliters in 1 liter.

milliliter (mL) (page 339)

This medicine dropper holds about 1 milliliter.

make a ten (page 55)

10 + 2

Make a ten to add 9 + 3.

minute (page 155)

1 minute

60 seconds = 1 minute

measure (page 317)

to find length, weight, capacity, or temperature

minute hand (page 155)

minute hand

Picture Glossary

missing addend (page 41)

9 + ☐ = 14

The missing addend is 5.

nickel (page 115)

5¢   5 cents

mode (page 509)

the number that occurs most often in a set of data

4   7   10   36   7   2

The mode is 7.

number line (page 19)

month (page 177)

This calendar shows the month of September.

September						
S	M	T	W	T	F	S
					1	2	3
4	5	6	7	8	9	10
11	12	13	14	15	16	17
18	19	20	21	22	23	24
25	26	27	28	29	30	

number sentence (page 253)

6 + 8 = 14 or 14 = 6 + 8

more likely (page 495)

Without looking, it is more likely that you will pick a ▪.

odd (page 105)

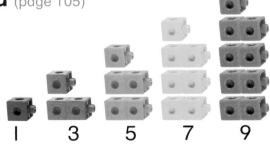

1   3   5   7   9

multiply (page 527)

2 × 4 = 8

ones (page 77)

3 ones

Picture Glossary

ordinal numbers (page 103)

numbers used to tell position

first   second   third

perimeter (page 379)

the distance around a shape

ounce (oz) (page 337)

One CD weighs about 1 ounce.

pictograph (page 195)

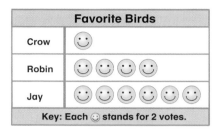

parallelogram (page 355)

4 sides and 4 angles

picture graph (page 189)

Our Favorite Pets	
Dog	🐶 🐶 🐶
Turtle	🐢
Cat	🐱 🐱 🐱 🐱

penny (page 115)

1¢   1 cent

pint (pt) (page 335)

2 cups = 1 pint

pentagon (page 355)

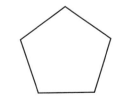

5 sides and 5 angles

place value (page 81)

the amount that each digit in a number stands for

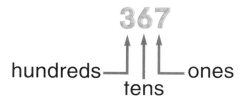

Picture Glossary

P.M. (page 171)

the hours from noon to midnight

It is 11:00 P.M.

quadrilateral (page 355)

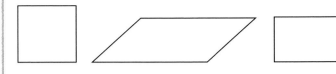

4 sides and 4 angles

pound (lb) (page 337)

The book weighs about 1 pound.

quart (qt) (page 335)

2 pints = 1 quart

prediction (page 497)

a telling that something will happen

quarter (page 121)

25¢   25 cents

probable (page 493)

an event that is more likely to happen

It is probable that you will pick a ■.

quarter hour (page 159)

quarter hour = 15 minutes

pyramid (page 353)

range (page 509)

the difference between the least number and the greatest number

4   7   10   36   7   2
              ↑   ↑
           greatest least

36 − 2 = 34. The range is 34.

LOG ON www.mmhmath.com to find out about these words

Picture Glossary

reasonable (page 259)

A reasonable answer makes sense.

19 + 32 = 51

20 + 30 = 50

51 is a reasonable answer.

round (page 259)

to find the ten or hundred closest to a number

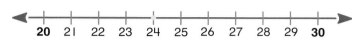

24 rounded to the nearest ten is 20.

rectangle (page 355)

4 sides and 4 angles

rule (page 21)

Add 3 is the rule.

Rule: Add 3

In	Out
10	13
20	23
30	33

rectangular prism (page 353)

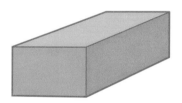

side (page 355)

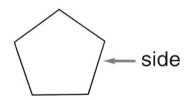

regroup (page 209)

12 ones = 1 ten 2 ones

skip-count (page 5)

Skip-count by 5.

related facts (page 37)

5 + 1 = 6

1 + 5 = 6

slide (page 377)

to move a figure horizontally, vertically, or diagonally

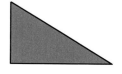
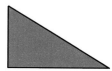

Picture Glossary

sphere (page 353)

square (page 355)

4 sides and 4 angles

subtract (−) (page 33)

5 − 3 = 2

sum (page 7)

sum
↓
3 + 2 = 5

The sum of 3 plus 2 is 5.

survey (page 191)

a way to collect data

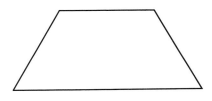

This survey shows favorite sports.

tally mark (page 191)

a mark used to record data

| = 1 |||| = 5

temperature (page 343)

a measure of hot or cold

The temperature is 79°F.

tens (page 77)

6 3
↑
6 tens

thousands (page 405)

1,253
↑
1 thousand

trapezoid (page 359)

4 sides and 4 angles

Picture Glossary

triangle (page 355)

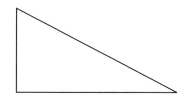

3 sides and 3 angles

week (page 177)

week →

There are 7 days in a week.

turn (page 377)

a figure that is rotated around a point

 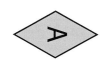

yard (yd) (page 321)

3 feet = 1 yard

unit (page 363)

A B A B A B A B

The things that repeat in a pattern make a unit.

year (page 177)

Jan	Feb	Mar	Apr
May	Jun	Jul	Aug
Sep	Oct	Nov	Dec

one year

vertex (page 353)

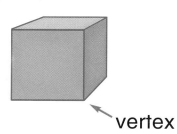

← vertex

Credits

Cover and title page photography: t. puppies-PhotoDisc; c. rug-Dot Box, Inc. for MMH.

Photography All photographs are by Macmillan/McGraw-Hill (MMH) and Peter Brandt for MMH except as noted below.

NA1: bkgd. Mark Barrett/Index Stock; b.c. Michael and Patricia Fogden/Minden Pictures; t.l. PhotoDisc; NA2: t.l. PhotoDisc; NA3: t.l. Richard Berenholtz/CORBIS; t.r. Don Cohen/DLC Photography; row 1 David A. Harvey/Getty Images; row 2 PhotoDisc; row 3 Pam Gardner/CORBIS. NA5: t.l. PhotoDisc; t.r. Annette Coolidge/PhotoEdit; b. PhotoDisc. NA6: PhotoDisc. NA7: l. PhotoDisc; r. Steve Chenn/Corbis. NA8: Cesar Lucas/Getty. NA9: PhotoDisc. NA10: t.l. John Conrad/CORBIS; t.r. Peter Harholdt/CORBIS; b.l. Joe McDonald/CORBIS; b.r. PhotoDisc. NA11: PhotoDisc. FL1: t. Richard Berenholtz/CORBIS; b. Brandon D. Cole/CORBIS; bkgd. John William Banagan/The ImageBank/Getty Images. FL2: Richard Berenholtz/CORBIS. FL3: t.l. Richard Berenholtz/CORBIS; t.r. Don Cohen/DLC Photography; row 1 David A. Harvey/Getty Images; row 2 PhotoDisc; row 3 Pam Gardner/CORBIS. FL5: t.l. Richard Berenholtz/CORBIS; t.r. Annette Coolidge/PhotoEdit; b. PhotoDisc . FL6: PhotoDisc. FL7: l. Richard Berenholtz/CORBIS; r. Steve Chenn/Corbis. FL8: Cesar Lucas/Getty Images. FL9: Richard Berenholtz/CORBIS. FL10: t.l. John Conrad/CORBIS; t.r. Peter Harholdt/CORBIS; b.l. Joe McDonald/CORBIS; b.r. PhotoDisc. NC1: t. Jim Schwabel/Index Stock; b. Joanne Pope/A Frame Or Two; bkgd. CORBIS. NC3: t.l. Jim Schwabel/Index Stock; t.r. Don Cohen/DLC Photography; row 1 David A. Harvey/Getty Images; row 2 PhotoDisc; row 3 Pam Gardner/CORBIS. NC5: t.l. Jim Schwabel/Index Stock; t.r. Galen Rowell/CORBIS; maple leaf Ron Watts/CORBIS; beech leaf Steve Satushek/Getty Images; oak leaf Ross M. Horwitz/Getty Images. NC6: birch leaf (t.l.) Martin B. Withers/CORBIS; hickory leaf (t.r.) Hal Horwitz/Getty Images. NC7: t.l. Jim Schwabel/Index Stock; t.r. Richard Hutchings/PhotoEdit; b. PhotoDisc. NC8: CORBIS. NC9: Jim Schwabel/Index Stock. NC10: l. CORBIS; c., r. PhotoDisc. ix: Mark E. Gibson/DRK Photo. xviii: l. Image Source/Image State; b. Roger De La Harpe/Animals Animals; bkgd. PhotoDisc/Getty images. 3: t., b. EYEwire. 25: r. Kennan Ward/CORBIS. 26: t. PhotoDisc; c.t. Ingram Publishing; c.b., b. PhotoDisc. 45: Karl Weatherly/PhotoDisc. 64: t. Martial Colomb/PhotoDisc; c.t. Siede Preis/PhotoDisc; c.b. PhotoSpin; b. Artville. 68A: Georgette Douwma/Getty Images; inset Comstock. 68B: t.l. Georgette Douwma/Getty Images; t.r. Comstock; b.r. Fotosearch; b.c. Fotosearch; b.r. Ryan McVay/Getty Images. 68C: t. Flip Nicklin/Minden Pictures; b.Flip Nicklin/Minden Pictures. 68D: t.l. Getty images; t.r. Getty Images; b. C Squared Studio/Getty Images; b.l.inset Georgette Douwma/Getty Images; b.r.inset Fotosearch.108: t. Digital Vision/Getty Images; c.t. C Squared Studio/PhotoDisc. 115: b.l. Artville. 116: c.t. Ingram Publishing; c.b., b. Artville. 119: t.l. Siede Preis/PhotoDisc; c.l. Artville; c.r. EYEWire. c.b. Corel; b. PhotoDisc. 128: CORBIS. 129: c.r. PhotoDisc c.l. Siede Preis/PhotoDisc; 139: t.l., c., b. PhotoDisc. 140: PhotoDisc. 142: t., b. Siede Preis/PhotoDisc; c.t. John A. Rizzo/PhotoDisc; c.b. Ingram Publishing. 145: b. PhotoDisc. 146A: Ariel Skelley. 146B: t. FotoSearch; b. Newscom. 146C: Joe Sohm/Chromosohm/Stock Connection/Picture Quest. 146D: t.l. Ariel Skelley; t.r. Newscom; b.l. FotoSearch; b.r.Joe Sohm/Chromosohm/Stock Connection/Picture Quest. 147: c. Craig Tuttle/Corbis; inset SuperStock. 148: c.l. John Shaw/Photo Researchers, Inc.; c.m.l. C Squared Studios/PhotoDisc; c.m.r. Jonelle Weaver/Getty Images; c.r. Kevin Schafer/CORBIS; b.l. John Shaw/Photo Researchers, Inc.; b.r. Jonelle Weaver/Getty Images. 156: b.c. Stephen Simpson/Getty Images; b.r. Ariel Skelly/CORBIS. 166: Jeff Zaruba/Getty Images. 173: c. C Squared Studios/PhotoDisc; b. Photos.com. 174: t. C Squared Studios/PhotoDisc; c. EYEWire. 186: Siede Preis/PhotoDisc. 201: Jan Halaska/Photo Researchers, Inc. 202: Mark E. Gibson/DRK Photo. 218: PhotoSpin. 222A: Peter Christopher/Masterfile; inset Jeff Greenberg/Index Stock. 222B: Ed Bock/CORBIS. 222C: r. Jeff Greenberg/Index Stock; b. Keith Brofsky/Getty Images. 222D: Mark Gibson/Index Stock. 223: c. Charles O'Rear/CORBIS; inset Astrid & Hanns Frieder-Michler/Photo Researchers, Inc. 263: Ann Purcell, Carl Purcell/Words & Pictures/PictureQuest. 264: t.c., b. PhotoDisc. 303: Tomas del Amo/Index Stock Imagery. 304: t., b. PhotoDisc. 308A: Keren Su/Getty Images; inset PhotoDisc. 308B: CORBIS. 308C: t. Scott McKinley/Getty Images; b. Daryl Balfour/Getty Images. 308D: t.l. Kennan Ward/CORBIS; t.r. Tom Brakefield/CORBIS; b. Chris Everard/Getty Images. 317: t.l. PhotoDisc; c.b., b. Artville. 318: t., t.c., b.c. Artville; c. David Toase/PhotoDisc. 320: c. Artville; b. Scott Harvey for MMH. 323: row 1, 3 Artville; row 2 StockByte; row 4 PhotoDisc. 324: c.t. Geostock/PhotoDisc; t.c.r. Siede Preis/PhotoDisc; l.c. C Squared Studios/PhotoDisc; b.c.r. PhotoDisc; b.r. Corel. 333: b.r. C Squared Studios/PhotoDisc. 334: b.r. Ryan McVay/PhotoDisc. 336: t.l. PhotoDisc; b.l. Burke/Triolo/Brand X Pictures/PictureQuest; b.r. John Campos/FoodPix. 337: t.l. PhotoSpin; t.r. Artville; b.r. Dot Box, Inc. for MMH. 338: t.l. PhotoDisc; c.r. EYEWire; b.l. Index Stock Imagery;. 339: c.r. PhotoSpin. 340: t.l., b.l. PhotoDisc; b.r. Index Stock Imagery. 341: t.l. Michael Matisse/PhotoDisc; b.l. Scott Harvey for MMH; b.r. EYEWire. 342: t.l. Dot Box, Inc. for MMH; t.c.l, b.c.l. PhotoDisc. 349: t.c. PhotoDisc; t.r. EYEWire; b.c., b.r. Ingram Publishing. 388A: Cover National Geographic; inset Siede Preis/PhotoDisc. 388B: b.l. Bettmann/CORBIS. 388C-b bkgd. Grant Heilman/Index Stock. 388C: r. Courtesy Jefferson National Parks Association. 388D: AP. 389: l. Michael Black/Bruce Coleman; l. inset Phil Degginger/Bruce Coleman. Inc.; r. M.Timothy O'Keefe/Bruce Coleman, Inc.; r. inset Deborah Davis/Photo Edit. 460: t. Stockbyte/PictureQuest; c.t. Siede Preis/PhotoDisc; c.b. Photos.com. 464A: cvr. SuperStock; inset PhotoDisc. 464B: Creatas. 464C: t. Brand X Pictures; b.l. Superstock; b.r. Comstock. 464D: t.l. Ablestock; t.c., t.r. PhotoDisc; b.l. C Squared Studio/Getty Images; c.l. Photospin; c.r. PhotoDisc; r.c.t., r.c.b., b.r. Ablestock. 465: c. Nancy Sefton/Photo Researchers, Inc.; inset Ken Karp for MMH. 466: Miep Van Damm/Masterfile. 506: PhotoDisc. 542A: cvr. G. Bliss/Masterfile; inset Bob Kris/CORBIS. 542B: t. Bob Kris/CORBIS; b. PhotoDisc. 542C: Rubberball. Martin B. Withers/CORBIS; hickory leaf (t.r.) Hal Horwitz/Getty Images.

Illustration Bernard Adnet: 473, 474. Farah Aria: 233, 234, 235, 236, 269, 431. Martha Aviles: 169. Kristen Barr: 195, 196, 523. Jennifer Beck-Harris: 58, 275, 276, 413, 415, 416, 451, 452. Shirley Beckes: 9, 10, 11, 14, 35, 36, 159, 173, 281, 282, 346, 421, 422, 475, 476, 481, 482, 487. Sarah Beise: 135, 137, 138, 403, 404, 469, 470. Carly Castillon: 25, 26, 117, 118, 152, 199, 207, 227, 239, 240, 241, 242, 297, 298, 361, 362, 401, 402, 435, 436, 493, 494, 500. Randy Cecil: 271, 272, 279, 280, 287, 293, 294, 447. Chi Chung: 21, 22, 200, 201. Nancy Coffelt: 5, 6, 63, 64, 299, 300, 477, 478, 507, 515. Diana Craft: 65, 66, 183, 265, 273, 274, 385. Lynn Cravath: 39, 40, 163, 164, 171, 172, 441, 442, 443. Mike Dammer: 2, 16, 32, 45, 52, 68, 76, 94, 114, 127, 132, 170, 174, 185, 188, 192, 203, 206, 208, 222, 230, 247, 252, 266, 292, 304, 314, 316, 324, 332, 352, 364, 367, 372, 376, 378, 384, 391, 396, 409, 410, 412, 414, 432, 448, 472, 495, 498, 506, 508, 524, 542, 543, 544, 545, 548, g4, g5, g8, g10, g11, g14. Nancy Davis: 193, 194, 204. Sarah Dillard: 81, 92, 103, 104. Kathi Ember: 95, 96, 165, 214, 215, 216, 253, 254, 405, 406, 489, 490. Buket Erdogan: 351. Dagmar Fehlau: 61, 62, 161, 162, 177, 178, 295, 296, 353. Brian Fujimoro: 179, 180, 251. Barry Gott: 41, 42, 85, 86, 89, 90, 99, 100, 191, 192, 305. Steve Haskamp: 83, 301, 302, 313, 314. Eileen Hine: 143. Tim Huhn: 33, 34, 97, 98, 259, 260, 417, 418, 423, 424, 457, 458. Melissa Iwai: 51, 197, 198, 231, 232. Jong Un Kim: 93. Richard Kolding: 87, 88. Erika LeBarre: 331, 538, 539. Chris Lench: 133, 134. Rosanne Litzinger: 261, 262. Lori Lohstoeter: 461, 462. Margeau Lucas: 15, 113. Lyn Martin: 53, 54, 109, 112, 483, 484, 531, 532. Christine Mau: 131. Debra Melmon: 157, 158. Laura Merer: 27, 517, 518, 547. Pat Meyers: 371. Edward Miller: 309, 310, 335, 337, 338, 341, 347, 368. Taia Morley: 343, 383, 519, 526, 528, 534. Keiko Motoyama: 175, 380. Christina Ong: 187. Laura Ovresat: 69, 70, 107, 108, 125, 127, 255, 256. Liz Pichon: sponsor critters, 31, 101, 102, 325, 326, 327, 328. Jen Rarey: 379. Mick Reid: 17, 18, 59, 60, 257, 258, 399, 400, 419, 501, 503. Stephanie Roth: 85, 86. Christine Schneider: 283, 284, 291, 491. David Sheldon: 73, 217, 245, 246, 313, 407, 408, 425, 433, 434, 444, 485, 486. Janet Skiles: 23, 24, 509, 510. Marsha Slomowitz: 119, 120, 243, 244. Ken Spengler: 219, 220, 499. Maribel Suarez: 355, 356, 449, 450. Susan Synarski: 7, 8, 12, 48, 189, 190, 285, 286, 459. Mary Thelen: 513, 514. Pamela Thomson: 3, 4, 43, 44, 181, 182, 365, 366, 427.

Acknowledgments

The publisher gratefully acknowledges permission to reprint the following copyrighted material:

Bushy-Tailed Mathematicians from COUNTING CATERPILLARS AND OTHER MATH POEMS by Betsy Franco. Copyright © 1998 by Betsy Franco. Published by Scholastic, Inc. Reprinted by permission.

Caterpillar Pete from POEMS TO COUNT ON by Sandra Liatsos. Copyright © 1995 by Sandra O. Liatsos. Published by Scholastic, Inc. Reprinted by permission.

Going To Bed. http://www.headstart.lane.or.us/education/activities/music/songs-fingerplays.html. Reprinted by permission.

Money Rhymes. http://www.canteach.ca/elementary/songspoems70.html. Reprinted by permission.

Skip-Count Song. http://www.canteach.ca/elementary/songspoems72.html. Reprinted by permission.

The Giraffe Graph from POEMS TO COUNT ON by Sandra Liatsos. Copyright © 1995 by Sandra O. Liatsos. Published by Scholastic, Inc. Reprinted by permission.

Twelve Little Rabbits. http://www.headstart.lane.or.us/education/activities/music/songs-fingerplays.html. Reprinted by permission.

The book covers listed below are reprinted with the permission of the following publishers:

Charlesbridge: THE COIN COUNTING BOOK

HarperCollins Publishers: GAME TIME!; IF YOU GIVE A PIG A PANCAKE; JUMP, KANGAROO, JUMP!; MISSING MITTENS; 100 SCHOOL DAYS; ONE LIGHTHOUSE, ONE MOON; ROOM FOR RIPLEY

HarperCollins UK: I SPY TWO EYES

Houghton Mifflin: SO MANY CATS!

Penguin Putnam Publishers: THE BUTTON BOX; HANNAH'S COLLECTIONS; TWENTY IS TOO MANY

Random House Publishers: HOW BIG IS A FOOT?

Scholastic, Inc. Publishers: THE CASE OF THE SHRUNKEN ALLOWANCE; THE GRAPES OF MATH

Book Cover for HOW MANY TEETH? reprinted by permission of the estate of Paul Galdone.

Book Cover for MOIRA'S BIRTHDAY by R. Munsch and M. Martchenko. Copyright © Michael Martchenko, artwork 1987. Reprinted with permission of Annick Press.

Book Cover for MRS. McTATS AND HER HOUSEFUL OF CATS reprinted with the permission of Margaret K. McElderry Books, an imprint of Simon & Schuster Children's Publishing Division, by Alyssa Satin Capucilli, illustrated by Joan Rankin. Illustrations copyright © 2001 Joan Rankin.

Book Cover for ROUND AND SQUARE by Miriam Schlein and Linda Bronson from Mondo's BOOKSHOP Literacy Program. Text copyright © 1999, 1952 by Miriam Schlein. Illustrations copyright © 1999 by Linda Bronson, reprinted by permission of Mondo Publishing, 980 Avenue of the Americas, New York, New York, 10018. All rights reserved.

Book Cover for WHAT'S NEXT, NINA? by Sue Kassirer, illustrated by Page Eastburn O'Rourke. Copyright © 2001 The Kane Press.